MIGHTY IS THE CHARM

Lectures on Science, Literature, and the Arts

J. Clifton Albergotti
University of San Francisco

. . . Mighty is the charm
Of those abstractions to a mind beset
With images, and haunted by herself,
And specially delightful unto me
Was that clear synthesis built up aloft
So gracefully. . . .

William Wordsworth
"The Prelude"

UNIVERSITY
PRESS OF
AMERICA

LANHAM • NEW YORK • LONDON

4720 Boston Way
Lanham, MD 20706

3 Henrietta Street
London WC2E 8LU England

Printed in the United States of America

Library of Congress Cataloging in Publication Data

Albergotti, J. Clifton
Mighty is the charm.

Includes bibliographies and index.
1. Science and the arts. 2. Science in art.
I. Title.
NX180.S3A4 700'.1.05 81-40158
ISBN 0-8191-2207-6 AACR2
ISBN 0-8191-2208-4 (pbk.)

Contents

Acknowledgements for Quotations

Page 2. From Galileo Galilei, *Two New Sciences,* translated by Stillman Drake (Madison: University of Wisconsin Press). Copyright © 1974 by University of Wisconsin Press. Reprinted by permission.

Pages 8, 9. From Werner Heisenberg, *Physics and Beyond* (New York: Harper and Row, Publishers, Inc.). Copyright © 1971 by Harper and Row. Reprinted by permission.

Page 17. From Reginald Bretnor, Ed., *Science Fiction, Today and Tomorrow* (New York: Harper and Row, Publishers, Inc.). Copyright © 1974 by Harper and Row. Reprinted by permission.

Page 17. From Brian Aldiss, *Billion Year Spree: The True History of Science Fiction* (New York: Doubleday and Company, Inc.). Copyright © 1973 by Brian Aldiss. Reprinted by permission of Doubleday and Company.

Page 17. From James O. Bailey, *Pilgrims Through Space and Time: Trends and Patterns in Scientific and Utopian Fiction* (Westport, CT: Greenwood Press). Copyright © 1947 James O. Bailey. Reprinted by permission of Mary M. Bailey.

Page 18. From Kingsley Amis, *New Maps of Hell* (London: Victor Gollancz, Ltd.). Copyright © 1960 by Kingsley Amis. Reprinted by permission of A. D. Peters and Co., Ltd.

Pages 18, 19. From Sam Moskowitz, *Explorers of the Infinite: Shapers of Science Fiction,* Classics of Science Fiction Series (Westport, CT: Hyperion Press, Inc.). Copyright © 1974 by Hyperion Press. Reprinted by permission.

Pages 24-26. From Johannes Kepler, *The Harmonies of the world,* translated by Charles G. Wallis, in Vol. 16, *Great Books of the Western World* (Chicago: Encyclopedia Britannica, Inc.). Copyright © 1952 Encyclopedia Britannica. Reprinted by permission.

Page 32. From Johannes Kepler, *Somnium,* translated by Edward Rosen (Madison: University of Wisconsin Press). Copyright © 1967 by University of Wisconsin Press. Reprinted by permission.

Page 70. From Alfred North Whitehead, *Science and the Modern World* (New York: Macmillan Publishing Company, Inc.). Copyright © 1925 by Macmillan, renewed 1953 by Evelyn Whitehead. Reprinted by permission.

Page 98. From Aristotle, *Aristotle's Poetics: A Translation and commentary for Students of Literature,* translated by Leon Golden, Commentary by O.B. hardison, Jr. (Englewood Cliffs, N.J.: Prentice-Hall, Inc.). Copyright © 1968 by Prentice-Hall. Reprinted by permission of Leon Golden and the University Presses of Florida.

Page 101. From Sir John Collings Squire, *Poems* (London: William Heinemann, Ltd.). Copyright © 1926. Reprinted by permission of W. Heinemann.

Pages 119-121. From Robert Frost, *The Poetry of Robert Frost* edited by Edward Connery Lathem. Copyright 1928, 1947, © 1969 by Holt, Rinehart and Winston. Copyright © 1956 by Robert Frost. Copyright © 1975 by Lesley Frost Ballantine. Reprinted by permission of Holt, Rinehart, and Winston, Publishers.

Pages 122-129. From Robinson Jeffers, *The Selected Poetry of Robinson Jeffers* (New York: Random House, Inc.). Copyright © 1959 by Random House. Reprinted by permission.

Pages 125, 126. From Sister Mary James Power, *Poets at Prayer* (London: Sheed and Ward, Ltd.). Copyright © 1938 by Sheed and Ward. Reprinted by permission.

Page 126. From Robinson Jeffers, *Be Angry at the Sun* (New York: Random House, Inc.). Copyright © 1941 by Random House. Reprinted by their permission and that of Donnan Jeffers.

Page 126, 127. From Robinson Jeffers, *The Beginning and the End* (New York: Random House, Inc.). Copyright © 1963 by Random House. Reprinted by permission.

Page 129. From Robinson Jeffers, *Hungerfield* (New York: Random House, Inc.). Copyright © 1954 by Random House. Reprinted by their permission and that of Donnan Jeffers.

Pages 131-134. From Kenneth Rexroth, *The Collected Shorter Poems of Kenneth Rexroth* (New York: New Directions Publishing Corp.). Copyright © 1966, 1963, 1962, 1952, 1949, 1940 by Kenneth Rexroth. Copyright © 1956, 1951, 1950, 1944 by New Directions. Reprinted by permission of New Directions Pub. Corp.

Page 135. From John Ciardi, *In Fact* (New Brunswick, N.J.: Rutgers University Press). Copyright © 1962 by John Ciardi. Reprinted by permission of the author.

Page 135. From John Ciardi, *As If* (New Brunswick, N.J.: Rutgers University Press). Copyright © 1955 by John Ciardi. Reprinted by permission of the author.

Page 136. From John Ciardi, *This Strangest Everything* (New Brunswick, N.J.: Rutgers University Press). Copyright © 1966 by John Ciardi. Reprinted by permission of the author.

Page 136. From John Ciardi, *39 Poems* (New Brunswick, N.J.: Rutgers University Press). Copyright © 1959 by John Ciardi. Reprinted by permission of the author.

Page 137. From John Updike, *Telephone Poles and Other Poems* (New York: Alfred A. Knopf, Inc.). Copyright © 1963 by Alfred A. Knopf. Reprinted by permission.

Page 191. From Isaac Newton, *Opticks,* Foreward by Albert Einstein (New York: Dover Publications, Inc.). Copyright © 1952 by Dover. Reprinted by permission.

Page 213. From Salvadore Dali, *Dali on Modern Art* (New York: The Dial Press). Copyright © 1957 The Dial Press. Reprinted by permission.

Pages 215, 216. From Aaron Scharf, *Art and Photography,* p. 151 (Harmondsworth: Penguin Books, Ltd.). Copyright © 1974 by Penguin Books. Reprinted by permission of Penguin Books, Ltd.

Page 221. From Jacques Hadamard, *The Psychology of Invention in the Mathematical Field* (Princeton: Princeton University Press). Copyright © 1945, 1973 by Princeton University Press. Reprinted by permission.

Acknowledgements for Illustrations

(The numbers in the following list refer to figures in the text.)

1	Drawing by Levin, © 1976 The New Yorker Magazine, Inc.
2, 3, 4, 5, 6, 9, 10, 15	The Bancroft Library, University of California, Berkeley
7, 17, 55	Stanford University Libraries, Department of Special Collections, Stanford, California
8	The Museum of Modern Art, Film Stills Archive, New York
11	Los Alamos Scientific Laboratory, Los Alamos, New Mexico
12	Courtesy of the Archives, California Institute of Technology, Pasadena, California
13, 16, 25, 26, 30, 32, 37, 44, 51, 52, 53, 54	Drawings by the author
14	National Portrait Gallery, London
18	The Hale Observatories, California Institute of Technology, Pasadena, California
19, 21, 22	Lick Observatory Photograph, University of California, Santa Cruz
20	Gleeson Library, Special Collections, Unviersity of San Francisco
23	Drawing by Charles Schulz, © 1969 United Feature Syndicate, Inc., New York
24	Photograph by Margo Moore, courtesy New Directions Publishing Corp., New York
27	Drawing by Mydung Thi Tran
28, 29	Steinway and Sons, New York
33, 34	The British Library, London
35, 36	The National Maritime Museum, Greenwich

38	Government of India Bureau of Tourism, Los Angeles
39	Cabinet des Estampes, Bibliothèque Nationale, Paris
40	Drawing by Randy T. Gouw
41, 42	The Kitt Peak National Observatory, Tucson, Arizona
43, 45	Rijksmuseum voor de Geschiedenis der Natuurwetenschappen, Leiden, The Netherlands
46, 50	Collection of the author
47	Museum of the History of Science, Oxford Photo: Edmark, Oxford
48, 49	Derby Borough Council Museums and Art Gallery, England

Preface

THAT AN AMERICAN university education should somehow include both science and the liberal arts as vital components in any curriculum is a proposition to which most educators give assent. To produce scientific literacy among liberal arts majors and to produce literate scientists among science majors seem to be two worthwhile goals. But what do the terms "scientific literacy" and "literate scientist" signify?

By "scientific literacy" I mean knowing some general things about science, such as its empirical nature, the interaction of theory and experiment, the creative role of the scientist, the tentative nature of the "laws" of science, and an appreciation of science as a human endeavor emerging from an historical context.

More elusive is an adequate definition of a "literate scientist." By this term I refer to a person whose education involves "the harmonious development of those qualities and faculties which characterize our humanity," as Coleridge said. The literate scientist is one who can manipulate the language of words as well as the language of symbols, who can make value judgements based on humanistic concerns, and who celebrates the scientific picture of the physical world as part of the human achievement of excellence which includes music, literature, and the arts.

Who are the literate scientists? Clearly, Freeman Dyson, Jacob Bronowski, and J. Robert Oppenheimer are shining examples. At the opposite end of the spectrum are those students (every university has its share) who are uninterested in anything beyond the narrowly practical.

A traditional way of promoting scientific literacy among nonscience majors is to require them to take a science course at some time during their academic career. This approach has two major shortcomings. First, the distribution in enrollment among the various science courses tends to be skewed in a predictable way—the courses requiring substantial mathematical background and significant amounts of homework are shunned by students. Second, if one tries to force the issue by narrowing the students' options to "hard-core" laboratory sciences, one still has the disadvantage of specialization. A non-science major who learns physics, for example, may achieve many of the goals of scientific literacy, as they apply to that particular discipline, but may feel hopelessly ignorant of other important scientific areas.

Another approach, teaching the history of science, avoids the specialization objection while aiding non-science majors in attaining the goal of scientific literacy. Science majors, whose knowledge of the history of their subject is often extremely sketchy, can also take a big step toward becoming literate scientists by a serious study of the history of science. Fortunately, history of science is becoming a more highly visible academic specialty, and scholars with these skills are becoming increasingly available.

The approach taken in this book—relating astronomy and physics to literature and the arts—has a strong historical component. If a student read no other text than this one, that student would at least learn something about the Scientific Revolution of the sixteenth and seventeenth centuries and the beginnings of the modern era. The "backbone" of the book is the often-told story of the development of celestial and terrestrial mechanics and the "successful" mathematical-physical method leading to the modern superbombs. (But Chapter 8 hints that other avenues, such as the development of optics and electro-magnetism, remain to be explored.)

The book is organized primarily around subjects—creativity, early science fiction, poetry, drama, and visual arts. When relevant to the subject matter discussed in the text, the scientific "heroes"—Kepler, Galileo, Newton, Oppenheimer, and Einstein—are singled out for "scientific profiles." Although these scientific biographies contain outlines of the major scientific achievements of each individual, I have kept in mind the reader with little or no scientific background. The book is roughly chronological, in that Kepler and Galileo are considered before Oppenheimer and Einstein. But it seemed more natural to treat Newton out of chronological order by discussing him in the chapter on English poetry where his literary influence could be readily documented.

A key word to characterize this book is "integrative." The scientific "heroes," flawed as they may be, have inspired a variety of poets, artists, composers, and even architects. A study of these related achievements integrates areas as seemingly disparate as physics and English poetry. Consider Galileo, for example. In order to appreciate Galileo more fully, one should study not only his scientific contributions and the dynamics of the interaction of seventeenth-century science and religion, but also his subsequent influence on drama, music, and poetry. Three playwrights have used the dramatic events of his life as the basis for biographical plays. One can trace in the poem "Paradise Lost" the residual effects of Milton's meeting with Galileo. A discussion of musical acoustics would be incomplete without mention of Galileo's equation relating the frequency of vibration of a stretched string to its tension. Finally, one should note John Ciardi's poem, "Galileo and the Laws." This holistic approach to Galileo aids the student in breaking down artificial compartmentalization of knowledge.

Since the choice of topics is inevitably a personal one, I apologize in advance for omitting someone's favorite poem, painting, or science-fiction work illustrating the interaction between science and the arts. The list can never be complete. I have taken astronomy and physics as examples of science *par excellence,* because they are the most mathematical of the sciences and I know them best.

When I have taught a course based on this book, one requirement has been a term paper or term project particularizing the course to the student's interest. The bibliographies at the end of each chapter are intended to be helpful in this regard. To each bibliography could be appended two references which have been extremely helpful to me in preparing this book: *The Dictionary of Scientific Biography,* Charles Coulston Gillispie, Editor-in-Chief (New York: Charles Scribner's Sons, 1970-1980), and the annual *Critical Bibliography* of *Isis,* the journal of the History of Science Society.

Thanks are due to the students who, over the years, have responded to my lectures with enthusiasm and with criticism. Thanks also to my colleagues Carolyn Merchant and Rodney Merrill for reading the book in manuscript form and making helpful comments. Finally, I wish to acknowledge the skillful typing and encouragement of Dean Hoffman.

Mill Valley
July, 1981

J.C.A.

Chapter 1

Creativity in Science and the Arts

IF ONE IS to judge from the ensuing controversy, C.P. Snow's book, *The Two Cultures* (1959), must have touched a raw nerve. The scholarly journals throughout the decade of the sixties were filled with articles which were in effect missiles in the warfare between the "literary intellectuals" (called by Snow, simply, "intellectuals") and the physical scientists, a conflict that had been simmering since the Industrial Revolution.

Rather than add directly to the barrage of words on this subject, I will merely observe that Snow himself, by virtue of being both a novelist and a scientist, bridged the gap between the two cultures. Similarly, one finds that Kepler combined the careers of astrologer, astronomer, and science-fiction author; that Galileo, the son of a composer, was accomplished on the lute in addition to being the father of two new sciences; that Wren was an astronomer as well as an architect; and that Tennyson was both an amateur astronomer and a poet. One could multiply the examples endlessly. These resonances between science and the arts will be explored in later chapters.

Each side in the "two cultures" controversy holds some mistaken notions of the other. Few "intellectuals" understand the true nature of science or of scientific research; on the other hand, many scientists have a naive conception of how artists work. After the false conceptions have been cleared away, then the true distinctions can emerge. To that end, this chapter compares the conventional Scientific Method, a simplistic and overly-mechanical description of how science works, with the Creative Process, the presumed *modus operandi* of musicians, artists, writers, and composers. First, I will examine what is missing from the usual outline of the Scientific Method. Then, I will review what the findings of modern brain research imply about the nature of the Creative Process. Finally, I will discuss the arguments concerning the alleged distinctions between scientific research and artistic creation. This analysis will suggest that the inception of scientific discovery has more in common with artistic creation than is generally supposed.

The Scientific Method

What is usually known as the Scientific Method was first formulated by Galileo, who conceived of a new way of doing scientific work. The process may be broken down into several, by now familiar, steps.

1. General observation of nature.
2. Hypothesis.
3. Deduction from hypothesis, or mathematical analysis.
4. Experimental test of the deduction. (Galileo placed special emphasis on this step.)
5. Revision of the hypothesis as a result of experimental test.

In his *Discourses on Two New Sciences,* Galileo, through the characters he created, comments on the significance of his scientific method—he calls it "a new contemplation"—and the future results to which it might lead.

> *Salviati:* Thus one may truly say that only now has the door been opened to a new contemplation, full of admirable conclusions, infinite in number, which in time to come will be able to put other minds to work.
>
> *Sagredo:* Truly, I believe that . . . these [principles of motion] which have been produced and demonstrated in this brief treatise, when they have passed into the hands of others of a speculative turn of mind, will become the path to many others, still more marvelous. This is likely to be the case because of the preeminence of this subject above all the rest of physics.

Newton acknowledged his great debt to Galileo 45 years after the death of the Italian pioneer when he wrote his *Principia,* a grand synthesis of terrestrial laws of motion with the movements of the heavenly bodies. He added a famous passage to the second edition of *Principia* which, at first glance, seems to denigrate Step 2 of the above Scientific Method. Newton said, *"Hypotheses non fingo"* (I frame no hypotheses). The context reveals, however, that Newton really meant that he would advance no hypotheses about the nature of gravity which were not subject to experimental verification.

> . . . Hitherto I have not been able to discover the cause of those properties of gravity from phenomena, and I frame no hypotheses; for whatever is not deduced from the phenomena is to be called hypothesis; and hypotheses, whether metaphysical or physical, whether of occult qualities or mechanical, have no place in experimental philosophy . . . And to us it is enough that gravity does really exist, and act according to the laws which we have explained, and abundantly serves to account for all the motions of the celestial bodies, and of our sea.

Privately, Newton suspected that an invisible fluid, the ether, transmitted the gravitational force across interplanetary distances. He did not print this speculation in the *Principia,* because at the time there was no way to test it experimentally.

Missing from the outline of steps of the Scientific Method are certain important elements. Some unstated assumptions or non-testable hypotheses (called "working hypotheses") are necessary just to get started. If they are stated, they are generally called "postulates." Prerequisite to science is an assumption that the phenomena of nature are ultimately explainable. In the western world, this belief has been more or less unconsciously absorbed as part of the Judeo-Christian world view. God created the universe; human beings who also stem from the original Divine creative act should be able to trace the Creator's steps. For God to leave false clues so as to deliberately deceive us would be out of character.

Another unstated belief is that nature is simple. Suppose one sets out to measure the spatial dependence of the gravitational force, and the best experimental results turn out to be $1/r^{2.0031}$. Now 2.0031 differs by a tiny amount, less than two-tenths of one percent, from the integer 2. The results of the experiment will undoubtedly be reported as $1/r^2$ plus or minus a certain experimental error, particularly since a theory exists which says that the force *ought* to be inverse-square. In this example, a belief in the simplicity of nature is reinforced by a physical theory giving the force a simple form. The criterion of simplicity applies as well to unnecessary proliferation of hypotheses. In this regard, scientists make liberal use of Ockham's Razor in the form, "What can be accounted for by fewer assumptions is explained in vain by more." Closely related to the criterion of simplicity is the requirement that a theory should be aesthetically pleasing. Copernicus rejected the Ptolemaic equants as being "not sufficiently pleasing to the mind." (The Ptolemaic and Copernican systems will be discussed in the next chapter.)

Finally, a belief in the universality of the laws of nature is not often stated. This deep-seated belief impelled Newton to search for the same mechanical explanation of motions of celestial bodies as that which applied to motions of bodies on earth. His Universal Theory of Gravitation succeeded where others had failed. Such concepts as celestial perfection and spirit propulsion were no longer needed.

The Scientific Method is not a formula which will automatically lead to success. The steps are for the most part straightforward, with the exception of Step 2. *How is the hypothesis to be obtained?* It may come from a desire for a beautiful or elegant mathematical statement. It may come from a modification of another hypothesis by chance or by analogy with a successful formulation, or it may stem from an inspired hunch. The genuinely creative elements which go into making a hypothesis are insufficiently publicized. Far more subjective or "human" elements are involved than the average person suspects.

The Creative Process

Artists, writers, composers, and other workers in the arts are commonly supposed to achieve their results not by the Scientific Method, but by the Creative Process. Graham Wallas, the English sociologist, has identified four stages in this process as follows:

1. Preparation
2. Incubation
3. Illumination
4. Verification

Not every creative person goes through all four stages, but most can identify several of these stages in their own creative process. Stage 3, Illumination, is the point at which creation takes place; it remains a mystery, even though it is placed in the midst of the more easily understood stages of Preparation, Incubation, and Verification. There are certain similarities to the Scientific Method, but in Wallas' characterization more emphasis is placed on the creative idea. Stage 1 is not unlike the fact-gathering of Step 1 in the Scientific Method, and may involve years of rigorous training and practice. The mathematician Henri Poincaré has compared this stage to the act of gathering together a collection of the hooked atoms of the Greek philosopher, Epicurus. These elements are hung on a wall where they remain motionless. During the Incubation stage, the mind removes these atoms from the wall, and begins to make innumerable combinations. Thus far a computer could take part in the creative process. Most of the combinations are not fruitful, but once in a great while a combination is recognized as beautiful and useful. This analogy emphasizes that the element of choice is important in the creative process, a discernment made according to aesthetic or other standards.

The swarming of the "atoms" may take place during a period of apparent restfulness or even unconsciousness, as Beethoven describes in the following letter:

Baden, Sept. 10, 1821

To Tobias von Haslinger
My Very Dear Friend,

On my way to Vienna yesterday, sleep overtook me in my carriage . . . while thus slumbering I dreamt that I had gone on a far journey, to no less a place than Syria, on to Judea and back, and then all the way to Arabia, when at length I actually arrived in Jerusalem . . . Now during my dream-journey, the following canon came into my head:

> [Here Beethoven writes several bars of music.]
>
> But scarcely did I awake when away flew the canon, and I could not recall any part of it. On returning here, however, next day, in the same carriage . . . I resumed my dream-journey, being on this occasion wide awake, when lo and behold! in accordance with the laws of association of ideas, the same canon flashed across me; so being now awake I held it fast as Menelaus did Proteus, only permitting it to be changed into three parts . . .
>
> (quoted in *The Creative Process,* B. Ghiselin, Ed.)

Incubation and Illumination stages are here telescoped into one. We may assume that he was semi-conscious, since the carriage motions probably did not allow a very deep state of sleep. The Verification stage is explicitly mentioned; he found it necessary to provide other parts for the canon. Significantly, he pushed the problem aside for a day, and then returned to it. Beethoven's inspirations did not always come in such flashes; an examination of his sketches for the last movement of the "Hammerklavier Sonata" shows a systematic testing and reworking of the theme of the fugue, evidence for "scientific" hard work rather than supposed "artistic" inspiration.

To take an example from literature, consider the case of Samuel Taylor Coleridge's account of the composition of "Kubla Khan."

> In the summer of the year 1797, the Author, then in ill health, had retired to a lonely farm-house between Porlock and Linton, on the Exmoor confines of Somerset and Devonshire. In consequence of a slight indisposition, an anodyne had been prescribed, from the effects of which he fell asleep in his chair at the moment that he was reading the following sentence, or words of the same substance, in 'Purchas' Pilgrimage': 'Here the Khan Kubla commanded a palace to be built, and a stately garden thereunto. And thus ten miles of fertile ground were inclosed with a wall.' The Author continued for about three hours in a profound sleep, at least of the external senses, during which time he had the most vivid confidence, that he could not have composed less than from two to three hundred lines; if that indeed can be called composition in which all the images rose up before him as *things,* with a parallel production of the corresponding expressions, without any sensation or consciousness of effort. On awaking he appeared to himself to have distinct recollection of the whole, and taking his pen, ink, and paper, instantly and eagerly wrote down the lines here preserved. (from "Prefatory Note to Kubla Khan")

In another account Coleridge indicates that the prescription he took was two grains of opium. Did the conscious self have to be stilled in order for the unconscious ideas to make themselves manifest? According to this view, the drug did not create the ideas, but merely induced a state of reverie in which free association could take place.

Fig. 1. The Creative Process in art. Drawing by Levin; ©1976 The New Yorker Magazine, Inc.

The Creative Act in Science

Thus far, the examples of the Creative Process have been drawn from the spheres of musical and literary composition, and it might be thought that the principles involved would be foreign to scientific work. But the contrary will be seen in Henri Poincaré's account of his discovery of an important class of mathematical functions. One need not understand the mathematical details in order to grasp the essential psychological truths Poincaré is trying to communicate. He relates the story in four paragraphs.

> For fifteen days I strove to prove that there could not be any functions like those I have since called Fuchsian functions. I was then very ignorant; every day I seated myself at my work table, stayed an hour or two, tried a great number of combinations and reached no results. One evening, contrary to my custom, I drank black coffee and could not sleep. Ideas rose in crowds; I felt them collide until pairs interlocked, so to speak, making a stable configuration. By the next morning I had established the existence of a class of Fuchsian functions, those which came from the hypergeometric series; I had only to write out the results, which took but a few hours.

In this extraordinary episode Poincaré achieves a kind of disembodied state during a sleepless night in which he contemplates the activities of his unconscious mind as if he were an outsider. His detached state was probably similar to Coleridge's opium-induced dream or Beethoven's productive slumber during a carriage ride. He next relates another development in the same mathematical theory.

> Then I wanted to represent these functions by the quotient of two series; this idea was perfectly conscious and deliberate, the analogy with elliptic functions guided me. I asked myself what properties these series must have if they existed, and I succeeded without difficulty in forming the series I have called theta-Fuchsian.

This time he achieves the result undramatically as in the classic

Scientific Method. In the final two instances Poincaré shows how the unconscious mind has been at work even during the time when the conscious mind has been occupied with diversions.

> Just at this time I left Caen where I was then living, to go on a geologic excursion under the auspices of the school of mines. The changes of travel made me forget my mathematical work. Having reached Coutances, we entered an omnibus to go to some place or other. At the moment when I put my foot on the step the idea came to me, without anything in my former thoughts seeming to have paved the way for it, that the transformations I had used to define the Fuchsian functions were identical with those of non-Euclidian geometry. I did not verify the idea; I should not have had time, as, upon taking my seat in the omnibus, I went on with a conversation already commenced, but I felt a perfect certainty. On my return to Caen, for conscience' sake, I verified the result at my leisure.
>
> Then I turned my attention to the study of some arithmetical questions apparently without much success and without a suspicion of any connection with my preceding researches. Disgusted with my failure, I went to spend a few days at the seaside, and thought of something else. One morning, walking on the bluff, the idea came to me, with just the same characteristics of brevity, suddenness and immediate certainty, that the arithmetic transformations of indeterminate ternary quadratic forms were identical with those of non-Euclidian geometry.
>
> ("Mathematical Creation" from *The Foundations of Science*)

These last two instances and numerous others testify to the value of temporarily putting one's work aside, and attacking it afresh at a later date. This insight is embodied in the popular expression, "Sleep on it." To use an electrical analogy, one may not be aware that a capacitor is charged with high voltage until it discharges suddenly when a conductor is placed across its terminals. Similarly, a period of intense intellectual effort "charges" the mind to a high "potential." It remains in this state while occupied with other matters (the Incubation Stage) until an apparently trivial stimulus, such as stepping into an omnibus or walking along a bluff, "discharges" it in a creative flash (the Illumination Stage).

Modern brain research into the separate functions of the two brain hemispheres (the "split brain") is beginning to throw some light on the nature of creativity. The left hemisphere receives sensory messages primarily from the right side of the body, and the right hemisphere primarily from the left side. This crossing of messages and mirror-image brain function we share with the other mammals. Unique to the human brain is a dichotomy of function as shown by Roger Sperry and collaborators at the California Institute of Technology. The left hemisphere in right-handers and most left-handers is specialized for language, analytical thinking such as mathematics and logic, and appreciation of temporal sequence of events. The right hemisphere, sometimes called the "silent hemisphere," has its own set of specialized abilities: music, body language, dreams, appreciation of spatial relationships and patterns, and intuition. The two modes of consciousness are most clearly revealed by the behavior of patients who have had the communication pathway between the two hemispheres (*corpus callosum*) severed as a treatment for severe epilepsy. In one experiment a composite face was shown to a patient so that the left half of the picture (half of a man's face) was seen by the left eye, and the right half of the picture (half of a gorilla's face) was seen by the right eye. If asked to talk (left-brain activity) about the picture, the patient would always report having seen a complete gorilla's face. When asked to point with his left hand (right-brain activity), the patient invariably would indicate a man's face. So each hemisphere not only had a different memory of the event, but also filled in the missing half of its picture.

What implications does this and similar research have for understanding creativity? The Incubation stage is evidently a right-brain activity. This hemisphere was at work while Beethoven was slumbering in his carriage, when Coleridge fell asleep in his chair, and during the swarming of Poincaré's "atoms." Not until a transfer of information across the *corpus callosum* occurs can one verbalize the discovery (the Illumination Stage). Can we develop simple, reliable exercises for the right hemisphere (the techniques of Yoga, Zen, the Sufis, or Christian meditation may be helpful here) just as we train the left hemisphere in our Western schools? To ignore the functions of the right hemisphere is to risk reducing our own creativity.

Poincaré realized that all mathematicians did not achieve their results in the same way that he did. He classified mathematicians as "intuitive" or "logical" (we might say primarily "right-brained" or "left-brained") according to whether their results were obtained in a flash of inspiration or as a result of logical, orderly develpment. Those in the "logical" category do not necessarily perform all of their work in a plodding, routine fashion. Moments of emotional and personal triumph occur during their work just as in the work of those in the "intuitive" category. I class the physicist Werner Heisenberg as one of the "logicals" in his development of quantum mechanics. He describes a night of feverish activity in which he worked out the new theory as follows:

> . . . one evening I reached the point where I was ready to determine the individual terms in the energy table, or, as we put it today, in the energy matrix, by what would now be considered an extremely clumsy series of calculations. When the first terms seemed to accord with the

energy principle, I became rather excited, and I began to make countless arithmetical errors. As a result, it was almost three o'clock in the morning before the final result of my computations lay before me. The energy principle had held for all the terms, and I could no longer doubt the mathematical consistency and coherence of the kind of quantum mechanics to which my calculations pointed. At first, I was deeply alarmed. I had the feeling that, through the surface of the atomic phenomena, I was looking at a strangely beautiful interior, and felt almost giddy at the thought that I now had to probe this wealth of mathematical structures nature had so generously spread out before me. I was far too excited to sleep, and so, as a new day dawned, I made for the southern tip of the island, where I had been longing to climb a rock jutting out into the sea. I did so now without too much trouble, and waited for the sun to rise.

(*Physics and Beyond*, p. 61)

The sense of exhaltation is unmistakable; at that moment Heisenberg was the only human being to know the secret of the atomic world which would shortly revolutionize physics.

The role of serendipity in the creative act must not be overlooked. The word does *not* come from a combination of serenity and stupidity as one of my students guessed! Horace Walpole coined the word in a letter (1751) he wrote to Horace Mann after reading the fairy tale, "The Three Princes of Serendip" (the ancient name of Ceylon). Walpole says in his letter, "As their highnesses traveled, they were always making discoveries by accident or sagacity, of things they were not in quest of." In the arts, the anonymous discoverer of egg tempura probably obtained the first mixture of egg yolk and pigment as a happy accident. The paintings of Jackson Pollack display a high degree of serendipity as the paint is dripped, splashed or brushed on.

One usually finds that a large amount of preparation precedes the typical accidental discovery. In science, take the familiar example of Charles Goodyear. His process of vulcanization of rubber stems from the accidental spillage of sulfur and India rubber on a hot stove (1893). Goodyear was in a unique position to take advantage of this accident, because he had worked for years trying to make rubber withstand the extremes of hot and cold. Two years prior to the accident he worked with a man who had made some experiments with mixtures of rubber and sulfur. The process was unperfected, but Goodyear optimistically bought the rights to it. Goodyear's experience bears out the saying of Pasteur, "In the fields of observation, chance favors only the minds which are prepared."

Procedures in Science and the Arts

If our critical examination of the creative act is shifted to the more general consideration of procedure, at least four additional similarities between the arts and sciences emerge.

Choice of a medium. A poet may choose to write an epic or a sonnet, for instance; a sculptor may use fiberglass, marble, or anodized aluminum. Similarly, a mathematician may work in differential geometry, set theory, or functional analysis; a physicist may formulate the laws of quantum mechanics in terms of differential equations or matrices.

Development of the subject matter in a unique style. A traditional subject may be treated conventionally or with originality. Many Dutch and Flemish painters of the 17th century depicted familiar scenes with meticulous detail and technical mastery, for example, Geraert Ter Borch's incredibly realistic treatment of satin or Aelbert Cuyp's bucolic scenes of placid cows. The "classical" (in the sense of being well understood) or traditional physics under active investigation today is, curiously enough, a relative newcomer, solid state physics. Essentially every phenomenon in this field of specialization has found a ready theoretical explanation, and predictions, such as room-temperature or organic superconductors (materials conducting electricity without resistance), are based on conservative extrapolations of today's knowledge. As an example of traditional subject matter being given a new twist, consider the artist Georges Braque, who painted a cubist interpretation of the still life. Similarly, Einstein showed that there was a very different way to look at traditional space and time.

Many creators reject traditional subject matter altogether, and allow their imaginations free rein. Abstract painters, such as Piet Mondrian, Wasily Kandinsky, and Franz Kline, fall into this category. Looking for particles whose existence has only been postulated is certainly an exploration of non-traditional subject matter in science. The imagination of physicists has led to a situation today where much money and talent is devoted to searching for quarks (particles with a fractional electric charge) and monopoles (isolated magnetic north or south poles). Such undertakings have in the past resulted in the discovery of numerous particles. The advocates of quarks and monopoles steadfastly hold to their predictions in the face of initial negative results. They have become modern Millerites (followers of the Baptist preacher who believed Christ's Second Coming would occur on October 22, 1844); if they, too, experience the Great Disappointment, their theories will suffer reappraisal.

Some slight restraints. A poet ordinarily uses words to communicate a poem, and a scientist ordinarily abides by the rules of logic. There are exceptions; a poet may coin new words or use nonsense syllables, and a scientist may venture beyond generalizations from observed facts in a bold guess.

Judgement of results. Several of the above unstated hypotheses may be applied in appraising the final product both in the arts and in the sciences. Does it have an inner simplicity and a universal character? Does repeated examination bring further insights? One cannot hope to discover all the secret depths of a great work at once. Thus, non-Euclidean geometry later served as a mathematical basis for Einstein's relativity theory, an application which was not at first suspected. Does the artist or scientist achieve a concentration of significance? A novelist searches for the *mot juste;* the language of mathematics is already a powerful and concise notation for thought. To epitomize Isaac Newton, one couplet of Alexander Pope suffices.

Questions of Uniqueness and Progress

Even if we observe certain similarities between the procedures of artistic workers and scientists, A.J. Ihde argues for an important distinction: a work of art is unique, whereas a work of science is inevitable. According to Ihde, if scientist **A** did not make his discovery today, then **B** would make it tomorrow. At first glance, the example of Einstein's Special Theory of Relativity seems to support this view. If Einstein had not published his Special Theory of Relativity when he did, others (H.A. Lorentz, H. Minkowski, and H. Weyl, for instance) might have discovered it later. But it might have been published piecemeal by several different authors, and it would not have had the same impact piecemeal as it had coming in its simplicity and profundity from the mind of Einstein. Thus Einstein's relativity theory was not inevitable.

Again, the independent discovery of the "four-factor formula" in nuclear reactor theory by six different groups in France, Germany, the Soviet Union, and the United States seems to support Ihde's case. The formula is closer to the more mundane aspects of engineering than to the lofty realms of relativity theory, limiting individual style and flair. Even within this narrow scope, as Spencer Weart has shown, chance circumstances, such as initial formulation of the problem, the particular bent of one or two key workers, and availability of materials, determine the direction the theoretical formulation will take and how it will be used in research. According to Weart's analysis, cultural differences are not as important as one might suppose. Thus even though simultaneous discovery occurred, we need not conclude that reactor physics developed inevitably according to a predetermined pattern.

Contrary to Ihde's view, a case could be made that works of art are not unique except in a very limited sense as in the example of Shakespeare's play, *Timon of Athens,* cited by Gunther Stent. Of course the exact word sequence is unique, just as in the case of Einstein's theory. More important is the content of the play. Stent asserts not only that the story of Timon could have been written without Shakespeare, but that it *was* written without him. He drew the plot from William Painter's *The Palace of Pleasure*. Painter had in turn used Plutarch and Lucian as sources. This points up the cumulative nature of the arts, a point often overlooked.

But if the story and character of Timon are not unique, then perhaps the deep insights are. Stent also questions this thesis. If Shakespeare had not existed, some other dramatist might have been able to probe the same depths of human emotion; consider that Shakespeare himself took the same theme, that of man's reponse to ingratitude, for his more successful *King Lear*. Thus we come to the same conclusion about Shakespeare's *Timon* as about Einstein's Special Theory of Relativity: these works are unique because no other dramatist or scientist could have written them in the same superlative way, although others might have been able to provide more or less the same insights.

Aldous Huxley believed another disparity exists between science and the arts (in this case literature). To him, the physical sciences are "nomothetic" (dealing with universals described by general laws), while literature is "ideographic" (concerned with individual or unique facts and processes). William Wordsworth, the English lake-country poet to whom nature, intensely felt, was a primary source of inspiration, would, on first inspection, seem to be a good example of the "ideographic" writer. The following lines from his poem, "The Tables Turned," describe the poet being seized by a numinous (i.e., "ideographic") experience while in the woods.

> One impulse from a vernal wood
> May teach you more of man,
> Of moral evil and of good,
> Than all the sages can.

But the poet could not share this unique event unless his experience struck a chord of universal ("nomothetic") resonance in his readers. Is Wordsworth talking about the particular forest he happened to be in? No, because that would make the poem inaccessible to, say, a resident of a tropical rain forest who had never seen an English woodland. The emphasis here is not on the particular forest, but on the universal human experience of awe in the presence of flora and fauna in a natural setting. What *is* specific is the language in which the universal experience is expressed. While the meanings are identical, the statement, "One sudden thought while walking through a spring forest . . .," lacks the power of the original Wordsworth utterance, "One impulse from a vernal wood . . . ." In another example from Wordsworth, the line, "The Child is father of the Man," from "My Heart Leaps Up . . .," is an expression having global validity. Even its form, that of an equation, expresses a general quality just as surely as Newton's law of force, F = MA. So the distinction between science as "nomothetic" and literature as "ideographic" is not as fundamental as Huxley supposed.

Another supposed difference between scientific research and artistic fabrication is that in art there is no progress; that is, one work of art is not superseded by another in the way that one scientific theory displaces another. According to this view there is change in art but no progress. Strong arguments bolster this viewpoint. A painting by Van Eyck, for example, remains a masterpiece, and a painting by Van Gogh, for example, does not supersede it. (I am ignoring temporary changes in popular taste which easily reverse themselves. For example, American collectors of the previous generation exceedingly prized the paintings of the English school, such as "Pinkie" and "Blue Boy" by Gainsborough; now their market value has declined in terms of real dollars. I suspect they will make a comeback. Popular taste fluctuations have less effect in science because scientists are judged by their peers, not the general public.) Progress in science is obvious if one uses a modestly priced pair of binoculars and considers that its optical quality is comparable to that of Galileo's telescope. Contrary to the argument that science progresses but art does not, Rossini's opera *Otello* (1816), based on the Shakespearean tragedy, was successful in its day, but Verdi's *Otello* (1887) has almost completely eclipsed its predecessor. Most critics would agree that Verdi's version has greater depth of characterization, and that a kind of musical progress has been achieved. Advocates of change in art/progress in science counter this by asserting that Rossini's opera was not a masterpiece: only bad art can be superseded.

As an alternative to the change in art/progress in science view, I suggest that a cyclical progression taking into account the gradual accumulation of cultural capital applies equally well to art and science. Consider the sequence of styles of poetry suggested by the names of John Milton, John Dryden, Alexander Pope, and William Wordsworth. Milton wrote in blank verse (unrhymed iambic pentameter), and Dryden wrote in rhymed couplets. Pope was a brilliant master of the rhymed couplet, and he used it in more subtle ways than Dryden. Pope favored understatement, Dryden forthrightness. Wordsworth's use of blank verse completes a cycle. When Wordsworth wrote blank verse, he and his readers were fully cognizant of what Milton had done with it. His verse was more personal and intimate in character than that of Milton. If a future poet revives the heroic couplet, that poet must face Pope's virtual exhaustion of the medium. Such cyclical progressions are also known in science: Newton espoused a particle theory of light which was superseded by Young's wave theory, only to be revived in a slightly altered form when Planck discovered photons or particles of light. Planck's photons are not identical with Newton's particles of light, but they incorporate them by reference. The current view is that light is more complicated than either model would suggest, and may manifest a wave or a particle nature, depending on the sort of experiment one devises.

In spite of the alleged differences between the Scientific Method and the Creative Process, scientific methodology does have room for subjective elements, aesthetic interests, and "human" qualities. Similarly, the "moment of inspiration" in artistic creation does not necessarily exclude logical planning and orderly development. Shall we call a truce in the battle of the "two cultures," and get on with the job of producing literate scientists and scientifically knowledgeable intellectuals?

Bibliography

Angus, D., "Quantum Physics and the Creative Mind," Amer. Scholar **30,** 212, 1961.

Beveridge, W.I.B., *The Art of Scientific Investigation* (W.W. Norton, New York, 1957).

Cannon, W.B., "The Role of Chance in Discovery," Sci. Monthly **50,** 204, 1940.

Coler, M.A., Ed., *Essays on Creativity in the Sciences* (New York Univ. Press, New York, 1963).

Garrett, A.B., *The Flash of Genius* (D. Van Nostrand, Princeton, 1963).

Ghiselin, Brewster, Ed., *The Creative Process* (Univ. of Calif. Press, Berkeley, 1954).

Hadamard, Jacques, *The Psychology of Invention in the Mathematical Field* (Dover, New York, 1954).

Heisenberg, Werner, *Physics and Beyond* (Harper and Row, New York, 1971).

Hurd, D.L., and Kipling, J.J., Eds., *The Origins and Growth of Physical Science,* Vols. 1 and 2 (Penguin Books, Baltimore, 1964).

Huxley, Aldous, *Literature and Science* (Harper and Row, New York, 1963).

Ihde, A.J., "The Inevitability of Scientific Discovery," Sci. Monthly **67,** 427, 1948.

Kuhn, Thomas K., "Energy Conservation as an Example of Simultaneous Discovery," in *Critical Problems in the History of Science,* M. Clagett, Ed. (Univ. of Wisconsin Press, Madison, 1969).

Libby, Walter, "The Scientific Imagination," Sci. Monthly **15,** *263, 1922.*

McLean, F.C., "The Happy Accident," Sci. Monthly **53,** 61, 1941.

Merton, Robert K., "Singletons and Multiplets in Scientific Discovery,"

Proc. Amer. Philos. Soc. **105,** 470, 1961.
Ornstein, Robert E., *The Psychology of Consciousness* (W.H. Freeman, San Francisco, 1972).
Roslansky, J.D., Ed., *Creativity* (North-Holland, Amsterdam, 1970).
Rothenberg, A., and Greenberg, B., *The Index of Scientific Writing on Creativity, 1566-1974* (Archon Books, Hamden, Conn., 1976).
Smith, Paul, Ed., *Creativity: An Examination of the Creative Process* (Hastings House, New York, 1959).
Snow, C.P., *The Two Cultures and A Second Look* (Cambridge University Press, 1969).
Stent, G.S., "Prematurity and Uniqueness in Scientific Discovery," Sci. Am. **227,** Dec., 84, 1972; **228,** March, 8, 1973.
Thomas, C.A., *Creativity in Science,* Arthur D. Little Memorial Lecture No. 8 (M.I.T. Press, Cambridge, 1955).
Utah, University of, *Research Conference on the Identification of Scientific Talent,* Conferences 1 through 3 (Univ. of Utah Press, Salt Lake City, 1959).
Weart, Spencer, "Secrecy, Simultaneous Discovery, and the Theory of Nuclear Reactors," Am. J. Phys. **45,** 1049, 1978.
Weaver, Warren, "Science and Imagination," Sci. Monthly **29,** 425, 1929.
Weisskopf, Victor F., "Is Physics Human?" Phys. Today **29,** 23, 1976.

Chapter 2

Precursors of Science Fiction

Towards a Definition of Science Fiction

To begin academically, one must attempt to define terms. What is *science fiction?* The two words are relatively clear when taken separately. *Fiction* is a prose narrative which treats imaginary situations. The narrative may be set in the past as in an historical novel, in the present as in a contemporary novel, or in the future as in many kinds of fantasy. *Science* is that body of knowledge based on sense data and organized according to mathematical or other conceptual schemes for the purpose of understanding ourselves and the nature of the world in which we live. When the two words are juxtaposed, complications arise, and no definition yet proposed is completely satisfactory. Some definitions I reject as being too all-encompassing. Consider the ones due to Bretnor and Aldiss.

> Fiction based on rational speculation regarding the human experience of science and its resultant technologies.
>
> (R. Bretnor in *Science Fiction Today and Tomorrow*)

> The search for a definition of man and his status in the universe which will stand in our advanced but confused state of knowledge (science), and is characteristically cast in the Gothic or post-Gothic mould.
>
> (B. Aldiss in *Billion Year Spress*)

Virtually any novel taking into account the effects of the Industrial Revolution, automation or television (to take examples at random) would qualify under these two criteria.

Avoiding the universality pitfall, J.O. Bailey gives the following brief definition:

> Narrative(s) of an imaginary invention or discovery in the natural sciences and consequent adventures and experiences.
>
> (*Pilgrims Through Space and Time*)

But his use of the word "imaginary" seems to rule out "real" inventions or discoveries. Surely in science fiction one can tolerate "real" space travel and Darwinian evolution, for example.

The social sciences would also appear to be off-limits according to Bailey's definition. My final objection is to the word "consequent," which makes science appear to be just a starting point for the action rather than an integral part of it.

More to my liking is a definition composed by Kingsley Amis.

> Science fiction is that class of prose narrative treating of a situation that could not arise in the world we know, but which is hypothesised on the basis of some innovation in science or technology, or pseudo-science or pseudo-technology, whether human or extra-terrestrial in origin. (*New Maps of Hell*)

Amis' definition excludes an historical novel dealing with scientists. Thus a novel such as *The Redemption of Tycho Brahe* by Max Brod, dealing with the conflict between Tycho Brahe and Kepler, does not qualify. The disqualifying factor is not its setting (after all, H.G. Wells often thrust his characters into the past in *The Time Machine*), but its lack of technical innovations to challenge the characters. Although readers generally allow a great deal of leeway, these innovations must be probable. Even the nearest star other than the sun, Alpha Centauri, is prohibitively far away for today's space ships. To cross our own galaxy at the speed of light would take over 100,000 years. Such problems may be solved by reasonable extrapolations of present-day science, such as frozen astronauts in suspended animation or Einstein time-dilation effects. Even strategems flatly contradicting today's science are sometimes accepted by readers. Material bodies cannot travel faster than the speed of light according to Einstein's theory, abundantly verified by experiments with atomic particles. This does not stop science fiction writers from putting their vehicles into "hyperdrive" or "space warp" (examples of Amis' pseudo-technology) in order to exceed the speed of light so that their characters might be more quickly ushered into the next arena of combat. So good science is not absolutely necessary for good fiction.

Although very helpful, Amis' definition is not without its problems. His definition excludes science-fiction *poetry*, a minor point, since few examples exist. Does his definition rule out "non-innovative" science? And what are we to make of the ambiguous phrase, "the world we know"? Numerous authors have engaged in straightforward extrapolation from the world of today to an anti-utopian future. Are these books to be excluded from the realm of science fiction?

The best definition of science fiction that I have found is one devised by Sam Moskowitz.

> A branch of fantasy identifiable by the fact that it eases the 'willing suspension of disbelief' . . . by utilizing an atmosphere of scientific

credibility for its imaginative speculation in physical science, space, time, social science, and philosophy.

(*Explorers of the Infinite*)

This formulation has the virtues of both breadth and specificity. Even so minimal requirement as "an atmosphere of scientific credibility" would seem to rule out Cyrano de Bergerac's *The Comical History of the States and Empires of the Worlds of the Moon and Sun.* The list of allowable speculations regrettably omits biological sciences. At the risk of offending Moskowitz, I will add biological sciences to the list, producing, in my opinion, a workable definition.

The literature of fantasy as opposed to science fiction generally deals with beings having magical powers, such as elves, witches, hobgoblins, and the like. No attempt is made to provide a factual basis for these powers. A good example is J.R.R. Tolkein's Ring trilogy, in which a moral framework and high imagination give heightened tension to a rousing adventure story. On the other hand, it must not be supposed that these creatures associated with folk tales and superstition can never appear in science fiction. They may arise in the mind of the protagonist as part of an alien hypnotic attack which weakens its victims by means of their own subconscious fears.

Although fantasy and science fiction are closely related, one can usually differentiate the two forms. The authors of science fiction place their characters in a matrix of scientific facts and emphasize logical explanations more than supernatural wonders, in other words, Moskowitz' "atmosphere of scientific credibility."

Science fiction has perhaps more in common with the detective story than with fantasy literature. Edgar Allan Poe, often called the father of the detective story, also wrote some stories ("The Facts in the Case of M. Valdemar," "Mellonta Tauta") that could be classed as science fiction. The two types of tales have in common an adherence to facts, an emphasis on action rather than characterization, and an intellectual puzzle painstakingly and logically worked out in the course of the narrative.

Authors who attempt to trace the history of science fiction often start with Lucian of Samosata (2nd century A.D.). A waterspout launches his travelers into the first interplanetary voyage; they find themselves on the moon eight days later. I shall bypass that and other early accounts in the ancestry of science fiction, and begin with Johannes Kepler's *Somnium* because of the author's pre-eminence as a scientist and his book's significance in the history of science fiction. For other early works of science fiction, readers may consult the excellent bibliography by R.M. Philmus in *Anatomy of Wonder* by Neil Barron (Ed.).

Johannes Kepler

Kepler's *Somnium* can be understood on the simplest level as a moon-journey narrative, one of the genre so well described by Marjorie Hope Nicolson in *Voyages to the Moon*. To discover his purposes for writing this science-fiction narrative, one must have an understanding of the four competing theories of the universe in the seventeenth century.

Diurnal Rotation of the Earth. This simple idea, that the earth was located in the center of the universe and rotated once a day relative to a stationary background of "fixed" stars, was held by Heraclides of Pontus (388 B.C.) and by later Phythagoreans, such as Hicetus of Syracuse. The planets were observed to wander in a complicated and unpredictable way among the stars. If a nova or supernova happened to appear, it was called a "guest" star. Comets were occasionally seen to move across the heavens. In most quarters, however, it was held that the earth was stationary, and that the heavens rotated once every twenty-four hours.

The Ptolemaic System. According to Greek ideas, the earth was the center about which rotated concentric spheres of some solid but transparent material in which the heavenly bodies were fixed. The number of spheres varied: 27 for Eudoxus, 34 for Callippus, and 55 for Aristotle. Eight spheres replaced the 55 of Aristotle in the period of Ptolemy (2nd century A.D.). From the earth outward air and fire were encountered, then in order came the spheres of the Moon, Mercury, Venus, Sun, Mars, Jupiter, Saturn, and the fixed stars. The eighth sphere containing the fixed stars was known as the Firmament because of its supposed function of steadying the spheres within. Outside the Firmament was a crystalline sphere added to account for the precession of the equinoxes, and a tenth, called the *Primum Mobile* (or first-moved) to set all the others in motion. Alexandrian astronomers, including Ptolemy, produced an astronomical encyclopedia, the *Almagest,* which synthesized the preceding Greek astronomy, and introduced some new concepts, effectively dominating astronomical thinking for 1300 years. Apollonius of Perga (third century B.C.) had envisioned each planet moving on a small circle called an *epicycle,* the center of which moved uniformly around a circle of large circumference known as the *deferent*. This combination of motions was needed to account for the observed retrograde motion (in which planets appeared to stop, move backward, and then forward in an open or closed loop) of the planets. Ptolemy introduced an imaginary point called the *equant* about which the deferent rotated. The equant was on a diameter of the deferent, and slightly displaced from the earth such that the equant and earth were on opposite sides of, and equidistant from, the center of the deferent. The result was a purely kinemat-

Fig. 2. Astronomia *watches over the Ptolemaic System. From* Margarita Philosophica *(1515) by Gregor Reisch.*

ical system which accounted for the observed motions of the planets, and enabled astronomers, after some calculations, to make predictions of future planetary positions.

The Copernican System. Nicolus Copernicus delayed publication of his book *De Revolutionibus Orbium Coelestium* until the end of his life, perhaps because he feared a negative reaction from ecclesiastical authorities. In his book Copernicus said that the planets revolved in uniform circular motion around the sun, and he retained about half the Ptolemaic epicycles. Galileo's telescopic observations strengthened this theory by analogy; Jupiter and its satellites were like a miniature solar system, and the earth was not unique in having a satellite. The phases of Venus were like the familiar phases of the moon, so it seemed reasonable that Venus orbited the sun just as the moon orbited the earth, effectively destroying the Ptolemaic geocentric argument. The result was an aesthetically superior, even elegant, formulation of planetary motions, but as yet no empirical proof of the earth's rotation or revolution existed.

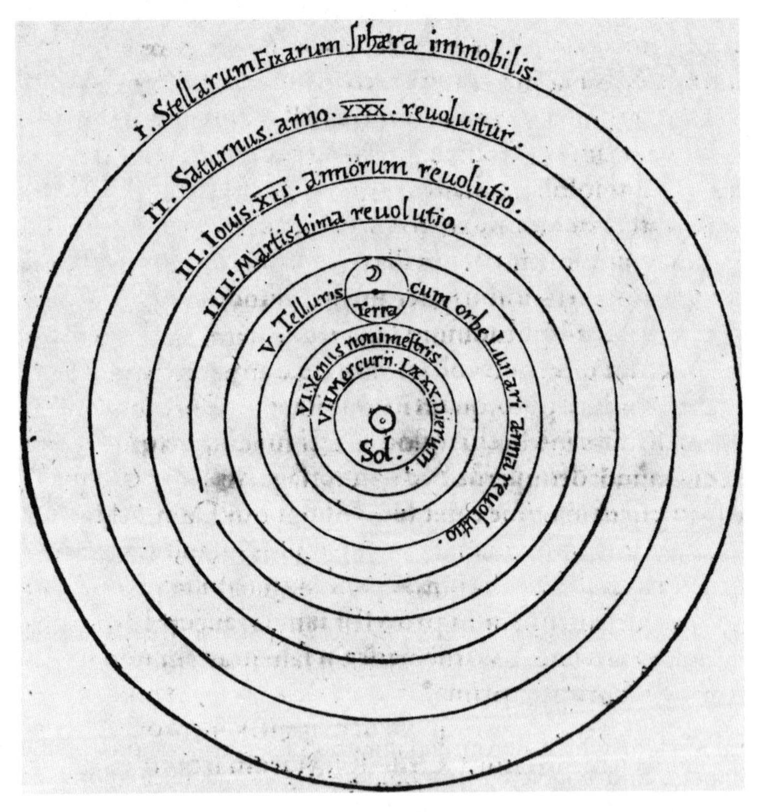

Fig. 3. The Copernican System from De Revolutionibus *(1543) by Nicolus Copernicus.*

The Tychonic System. The Danish nobleman Tycho Brahe is generally recognized as the greatest observational astronomer before the invention of the telescope. He retained some details of the Ptolemaic scheme; for example, he placed the earth in the center of the universe with the sphere of fixed stars rotating about the earth once every twenty-four hours. The moon circled the earth, and so did the sun, but all the other planets circled the sun. Mathematically, Tycho's conception is identical to the Copernican one. A deciding factor between the two rival theories would have been the difference in apparent stellar motions predicted by the two theories, but definitive measurements of stellar parallax were not available until 1838.

After Brahe's death, his heirs eventually transferred his observational records to Kepler, his assistant, who painstakingly extracted from them the basis for his three empirical laws of planetary motion which significantly modify the Copernican scheme by doing away with epicycles and circular orbits. 1) *The law of ellipses.* The orbit of each planet is an ellipse with the sun at one focus.[1] 2) *The law of equal areas.* The radius vector from sun to planet sweeps out equal areas in equal times. 3) *The harmonic law.* The cube of the mean distance from sun to planet is proportional to the square of the period of revolution. From these laws Kepler was able to calculate the positions of the planets with great accuracy as published (1627) in his *Rudolphine Tables* (named in honor of his patron, Holy Roman Emperor Rudolph II).

Kepler's three laws are of sufficient generality to describe any planetary system in which the members move independently of one another, but he was never able to fathom the physical principles underlying the astonishing regularities he had discovered. Kepler's agony over abandoning circular orbits, and his unbounded joy in discovering the harmonic law deserve further scrutiny, because Kepler in his own person represents a bridge between the two cultures of C.P. Snow, between astronomy and astrology, in a way that has not been possible since.

A Scientific Profile. In 1589 Kepler took a theological scholarship to the University of Tubingen, intending to become a Lutheran minister. Although he gladly accepted a position in mathematics when it opened up in Graz, he never completely submerged his theological and mystical bent. He was deeply committed to the ideas of the Neo-Platonists who believed that images or archetypes given by God pre-existed in the human

[1]Kepler was the first to use the term "focus" (from the Latin for fireplace or hearth), but he used it in connection with his optical studies for the place where a lens brings together parallel rays of light from a distant source. He was also the first to notice that moonlight concentrated by a spherical mirror will cause heating like a "certain warm breath."

psyche. The task of the scientist was to match these ideal forms to the surrounding universe. Kepler employs the same word as the psychologist, C.G. Jung, uses for his archetypes or "primordial images." One such ideal form to which Kepler attached supreme importance was the sphere, to him a symbol of the Trinity. The analogy went as follows: the central point represented God the Father, the surface God the Son (the face the sphere presented to the outside world), and the radii from center to surface God the Holy Spirit. The sphere would cease to exist if any one of these three components were to be removed. A plane intersecting a sphere determines a circle, the Trinitarian analog of two-dimensional space. Since an ellipse lacked these associations in Kepler's mind, for him to abandon his favorite conception, the circle, in the first law was a considerable sacrifice.

Kepler could not have arrived at his first law had he not obtained Tycho Brahe's data, the most accurate (about 2 minutes of arc for stars, 4 minutes for planets) available at the time. Brahe made his observations without benefit of telescope from the Uraniborg Observatory on the island of Hven in Denmark. If plotted accurately to scale, most planetary orbits have so little eccentricity that they would appear at first glance to be circles. If Brahe's data had been less accurate, the orbital eccentricity would not have been noticeable, but if the data had been more accurate, the eccentricity might have been obscured by perturbations due to the presence of nearby planets.

Of the planets known to Kepler, Mercury had the greatest eccentricity but was hard to observe because of its proximity to the sun. Mars had the next greatest eccentricity. Kepler launched what he called his "war on Mars" to determine its orbit, not a trivial problem since the shape of the earth's orbit was unknown, nor did the earth provide a central position for observing the motion of Mars. After trying successively to fit the data to a circle and an oval, he decided that the best fit was to an ellipse with the sun at one focus.

Kepler's second law implied that the closer a planet was to the sun the faster it moved, generating equal areas within its orbit during equal time intervals. This variation in velocity disturbed his sense of the way in which God would have set the universe in motion.

With the third law he finally attained a relation harmonizing with his Neo-Platonic ideas, because the powers 2 and 3 appearing in this law are small whole numbers. With regard to this discovery he writes triumphantly in *Harmonies of the World* (1619) as follows:

> But now since the first light eight months ago, since broad day three months ago, and since the sun of my wonderful speculation has

Fig. 4. Tycho Brahe using his great quadrant at Uraniborg, Denmark. From Astronomiae Instauratae Mechanica *(1602) by Tycho Brahe.*

shown fully a very few days ago: nothing holds me back. I am free to give myself up to the sacred madness, I am free to taunt mortals with the frank confession that I am stealing the golden vessels of the Egyptians, in order to build of them a temple for my God, far from the territory of Egypt. If you pardon me, I shall rejoice; if you are enraged, I shall bear up. The die is cast, and I am writing the book—whether to be read by my contemporaries or by posterity matters not.

> Let it await its reader for a hundred years, as God Himself has been ready of His contemplation for six thousand years.

Besides astronomy, Kepler was interested in geometry and music. The famous third law is merely the eighth of thirteen theorems in the fifth book of *Harmonies of the World*. He needed this theorem for something dear to him, proof that musical harmonies underlie planetary motions. He felt that harmonic relationships took precedence over simple geometry and quantity.

> . . . the consideration of priority and harmonic perfection comes first, and the consideration of quantity comes last, because there is no beauty in quantity of itself.

(This passage from *Harmonies of the World* illustrates the non-testable hypothesis of Chapter 1.)

According to the notion of the "music of the spheres,"[2] an idea from the ancient Greeks, each planet emits a sound, and the higher the speed of the planet, the higher the pitch of its associated sound. Shakespeare refers to this in the speech of Lorenzo to Jessica in *The Merchant of Venice*.

> Sit, Jessica. Look how the floor of heaven
> Is thick inlaid with patines of bright gold;
> There's not the smallest orb which thou behold'st
> But in his motion like an angel sings,
> Still quiring to the young-eyed cherubins;
> Such harmony is in immortal souls;
> But whilst this muddy vesture of decay
> Doth grossly close it in, we cannot hear it. (Act V, Scene 1)

Kepler did not actually expect to hear the planets, since their voices were audible only to the ear of God. After discarding a great variety of numbers associated with the planetary motions, he at last discovered that the velocities of the planets at aphelion and perihelion form ratios of small whole numbers. These ratios can easily be expressed as distances on a monochord, and interpreted as musical harmonics as indicated by the last column in Table 1. He was then able to arrange the orbital speeds of a planet in a sequence of notes from the lowest to highest (see Fig. 5), thus forming a signature tune for each planet. Finally, he combined the tune for the various planets into a cosmic chorus, with bass, baritone, alto, and soprano parts. Kepler's joy over these discoveries knew no bounds. For him, it made up for the loss of the

[2]The "music of the spheres" surfaces in 20th-century science fiction in C.S. Lewis' *That Hideous Strength*. When the planetary gods descend on the house at St. Anne's, the inhabitants experience "melody," "rhythm of such fierce speed," "a marching-song," "ordered rhythm of the universe," and "the *Gloria* which those five excellent Natures perpetually sing."

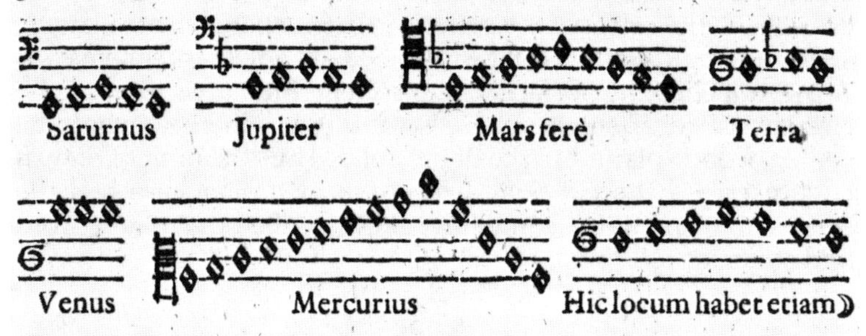
HARMONICIS LIB. V. 207

mnia (infinita in potentiâ) permeantes actu : id quod aliter à me non potuit exprimi, quam per continuam ſeriem Notarum intermedia-

rum. Venus ferè manet in uniſono non æquans tenſionis amplitudine vel minimum ex concinnis intervallis.

Fig. 5. The music of the spheres from Kepler's Harmonices Mundi V *(1619).*

Table 1

Planets		*Ratio of Distances*	*Musical Terminology*
Saturn	aphelion a	a/b = 4/5	Major third
	perihelion b	a/d = 1/3	Twelve tone
		c/d = 5/6	Minor third
Jupiter	aphelion c	b/c = 1/2	Octave
	perihelion d	c/f = 1/8	Three octaves
		e/f = 2/3	Perfect fifth
Mars	aphelion e	d/e = 5/24	Minor third plus two octaves
	perihelion f	e/h = 5/12	Minor third plus one octave
Earth	aphelion g	g/h = 15/16	Diatonic semitone
	perihelion h	f/g = 2/3	Perfect fifth
		g/k = 3/5	Major sixth
Venus	aphelion i	i/k = 24/25	Chromatic semitone
	perihelion k	h/i = 5/8	Minor sixth
		i/m = 1/4	Double octave
Mercury	aphelion 1	l/m = 5/12	Minor third plus one octave
	perihelion m	k/l = 3/5	Major sixth

cherished circle in the first law.

In an earlier discovery of which Kepler was inordinately proud, he related the mean distances of the six known planets to the five regular solids. Because of this achievement, as a young man of 23, he came to the attention of Tycho Brahe. The last of the thirteen books of Euclid contains the theorem that there are only five regular polyhedra, the culminating achievement of ancient geometry. Kepler attached great significance to the existence of five and only five of these objects, and arranged them as a nest of boxes with the orbits of the planets in between. Since the Platonic solids are regular, they can be inscribed in a sphere such that each corner touches the surface of the sphere. Similarly, they can be circumscribed about a sphere so that the sphere touches the center of each plane surface of the solid. Thus the orbit of Saturn circumscribed the cube, and the orbit of Jupiter inscribed the same cube, and at the same time circumscribed the tetrahedron, and so on, as follows:

Orbit of Saturn
Cube
Orbit of Jupiter
Tetrahedron
Orbit of Mars
Dodecahedron
Orbit of Earth
Icosahedron
Orbit of Venus
Octahedron
Orbit of Mercury

The ratio between radii of inscribed and circumscribed spheres is given by the following simple formula (S.M. Coxeter, *Introduction to Geometry*) which can be easily worked out on an electronic calculator:

$$\frac{R\,(\text{inscribed})}{R\,(\text{circumscribed})} = \cot\frac{\pi}{p}\quad\cot\frac{\pi}{q},$$

where p is the number of edges in any face of the solid, q is the number of faces situated about any vortex (for the dodecahedron, for example, p is 5 and q is 3), and the calculation must be carried out in radians rather than degrees. Table 2 shows the results. If one allows the calculation based on the octahedron to use the circle inscribed in the square base of the hemi-octahedron rather than the inscribed sphere, the fit with the observations of Copernicus is surprisingly good. Is this science or is it mysticism and numerology? This unification of astronomy and geometry was deeply satisfying to Kepler; he felt certain that he had traced the mind of the Creator in this matter.

An alternative way of expressing the regularities Kepler worked out with his nest of boxes is the relation announced by Johann Daniel Titius (1772) and publicized by J.E. Bode. The Titius-Bode law, remarkable for its simplicity, is as follows:

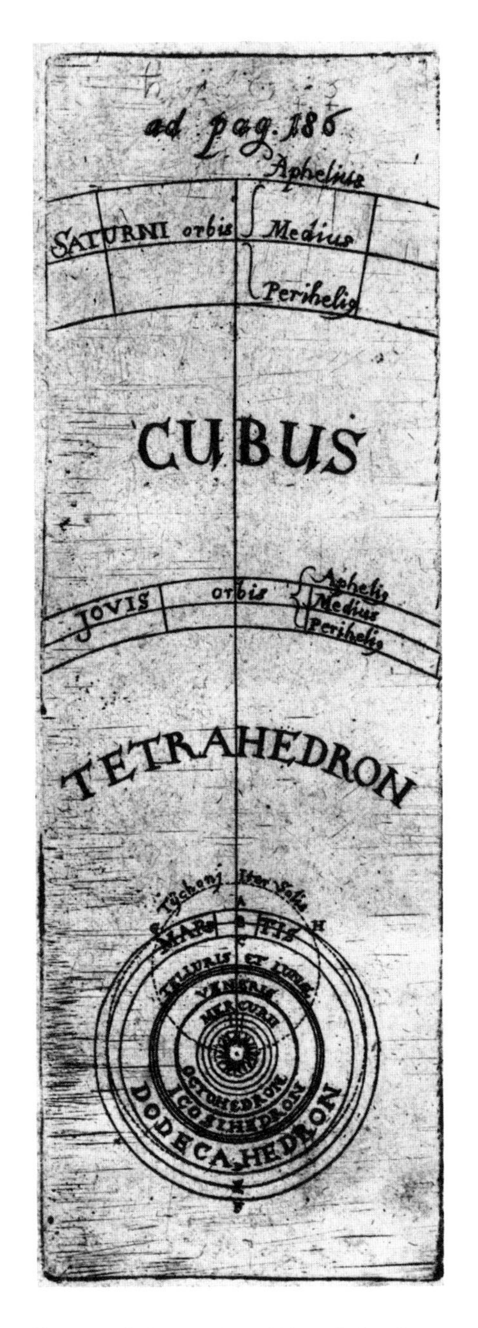

Fig. 6. Kepler's diagram of the regular solids and the planetary orbits. From Mysterium Cosmographicum *(1621) by Johannes Kepler.*

Table 2.

Ratios of Average Planetary Distances Based on the Platonic Solids

Planet	Regular Polyhedron	Theoretical Ratio (thin shell)	Observed Ratio (based on data of Copernicus)
Mercury/Venus	Octahedron	0.707	0.723
Venus/Earth	Icosahedron	0.795	0.794
Earth/Mars	Dodecahedron	0.795	0.757
Mars/Jupiter	Tetrahedron	0.333	0.333
Jupiter/Saturn	Cube	0.577	0.635

$$R \propto 4 + (3) \times (2)^{i-2}$$

For Venus i is 2, for Earth 3, for Mars 4, for Jupiter 6, and for Saturn 7. Unfortunately, Mercury does not fit for i = 1 as seems natural, but the formula works if i is set equal to minus infinity for that planet. Uranus was discovered by William Hershel in 1781, and its distance fits the scheme with a value of 8 for the power i. The gap at i = 5 was neatly filled by the discovery of Ceres in 1801, the first member of the asteroid belt to be seen. The formula fails badly for Neptune and Pluto, although the fit is better if one treats Neptune and Pluto as one planet, on the theory that Pluto was formerly a satellite of Neptune. For about 200 years the Titius-Bode law was a mystery until some recent computer simulations of solar system formation and subsequent planetary dynamics began to shed some light on the problem. Why does this approach seem more satisfactory to the modern scientific mind than Kepler's?

Kepler revised his original polyhedral construction to allow each sphere containing an orbit of a planet to have a certain thickness. Table 3 compares modern values of planetary distances with those computed with the aid of Kepler's construction and Bode's law. For the Keplerian values the thickness of each shell has been adjusted according to eccentricity of orbit, or natural satellite orbit diameter, whichever is largest. The agreement between the modern values and the planetary distances obtained from Bode's law is remarkably good except for the outermost planets.

Kepler's correlation of the orbits of the six Copernican planets with the five Platonic solids and his elucidation of the music of the spheres—"queer stuff," in the words of I. Bernard Cohen—are inconsistent with the modern concept of scientific research. But Kepler himself was just as proud of these discoveries as he was of his three laws of planetary motion upon which his fame mainly rests today. If the soundness of an astronomical system depends upon its ability to make predictions of the future positions of the observable celestial bodies, then Kepler's astronomical work has been fully vindicated. His three decades of thought and computation embodied in the *Rudolphine Tables* enabled him to predict that Mercury would traverse the sun's disc on 7 November 1631. Less than a year after Kepler's death, the transit was observed within five hours of the predicted time. A fuller discussion of Kepler's place in the history of science would have to include not just astronomy, but optics, physiology and psychology of perception, pure and applied mathematics, and crystallography.

If Kepler had restricted his imagination to the observations of planetary positions made from earth, the idea of elliptical planetary orbits as described in his first law probably would not have occurred to him. In order to visualize an elliptical orbit, he needed an extra-terrestrial viewpoint. In his fictional work *Somnium,* which will be considered next, he lifts his characters to the moon, a convenient observation platform for viewing a variety of celestial phenomena.

Kepler's Science-Fiction Narrative. *Somnium* or *Dream* (posthumously published in 1634) is a work of science fiction written to popularize a scientific theory. Modern counterparts are relatively rare; George Gamow's *Mr. Tompkins in Wonderland* is a populari-

Table 3.

Average Distances of Planets from the Sun (in Astronomical Units)

Planet	Modern Value	Bode's Law	Kepler's Polyhedral Construction (thick shell)
Mercury	0.39	0.40	0.45
Venus	0.72	0.70	0.79
Earth	1.00	1.00	1.00
Mars	1.52	1.60	1.26
Asteroid Belt (Ceres)	2.77	2.80	—
Jupiter	5.20	5.20	4.0
Saturn	9.55	10.0	7.2
Uranus	19.19	19.6	—
Neptune	30.07	38.8	—
Pluto	39.5	77.2	—

zation of Einstein's Special Theory of Relativity, and James Gunn's *The Listeners* attempts to rationalize the search for extraterrestrial life. The scientific theory Kepler wished to promote was the Copernican heliocentric hypothesis, because his own work was a modification of it. In a note to *Somnium* he says,

> Everybody screams that the motion of the heavenly bodies around the earth and the motionlessness of the earth are manifest to the eyes. To the eyes of the lunarians, I reply, it is manifest that our earth, their Volva, rotates, but their moon is motionless. If it be argued that the lunatic senses of my lunarian people are deceived, with equal right I answer that the terrestrial senses of the earth dwellers are devoid of reason. (Note 146, *Somnium*)

The work consists of four parts: the narrative proper, a set of notes about three times longer than the narrative, a *Geographical or If You Prefer, Selenographical Appendix,* and a set of notes on the appendix. The first part of the narrative has many autobiographical details for which Kepler was to pay dearly, as will be seen. The tale is set in Iceland, and describes Duracotus and his mother Fiolxhilde, a "wise woman" who sells little bags of herbs with magical powers. After Duracotus rips open a bag of herbs which had been purchased by a ship's captain, Fiolxhilde impulsively offers her son to the captain instead of the damaged bag so that she can retain the captain's money. After the captain deposits him on the coast of Denmark with instructions to deliver a letter to Tycho Brahe, he becomes Brahe's student. (Kepler himself was an assistant of Brahe, not a pupil, as Duracotus is in *Somnium.*) After several years Duracotus returns to Iceland, rejoicing to learn that his mother is also astrologically knowledgeable. In an occult ceremony she summons a demon from Levania (the moon) who describes a trip to his homeland.

The work was published posthumously, but manuscript versions circulated prior to Kepler's death, and through incredible misunderstanding and deliberate distortion, were used to increase prejudice against his mother after she had been brought to trial for witchcraft on other charges. Kepler, coming to her aid, found her chained in a dungeon. During the trial, which lasted a year, Kepler argued, not that witches did not exist, but that his mother had never committed acts which would make her one. She was acquitted, but died a year after release from prison.

The voyage to the moon in *Somnium* is accomplished by "demon power." This mode of transport might disqualify the story as science fiction, but the author embellishes the trip with so much scientific detail that, on balance, the account is more scientific than supernatural. The mixture of supernatural and scientific elements is typical of Kepler, who stands on an intellectual continental divide between the Middle Ages and the Renaissance.

During the journey to the moon (four hours—the approximate duration of a solar eclipse) demons can freely move within the cone of the earth's shadow carrying human beings with them. Travelers on this first modern scientific moon-voyage must cope with a number of hardships. Since the journey must necessarily be swift, Kepler recommends opiates to relieve the pains the travelers would otherwise suffer. He recognizes that they will encounter extreme cold as well as lack of air. Applying damp sponges to the nostrils alleviates the breathing difficulty.

Somnium also contains Kepler's speculations on the dynamics of the journey. Aristotle had said that celestial bodies had a tendency to revolve forever. In Kepler's concept of inertia, celestial bodies have a propensity for rest or quiet, from which they must be moved by a force. His favorite was the magnetic force, and he recognized that at some point between earth and moon their oppositely directed magnetic attractions would be equal, exerting zero force upon the traveler. At this point in the narrative he indicates that a mass would proceed of its own accord, but in the notes he fails to make this explicit.

Upon arrival the travelers find the moon's surface to be similar to the descriptions of Galileo's telescopic observations of 1610, but Kepler postulates water on the moon in opposition to Galileo's views. The travelers notice that the surface of the moon is porous, with many caves and grottoes, affording the inhabitants refuge from the extremes of heat and cold which they would otherwise suffer on the surface. No human beings exist there, but creatures mostly of "serpentine nature" which fly, walk, or swim. Rapid vegetative growth, a theme often exploited in subsequent science fiction, causes plants to reach prodigious size in one lunar day, 15 earth days in duration. Mountains on the moon exceed in height anything found upon earth. Circular moon craters are taken by Kepler to be fortifications constructed by the inhabitants using a cord tethered to a point.

The main body of *Somnium* is a detailed description, for the most part accurate, of a multitude of celestial phenomena seen from the viewpoint of the moon. Kepler has used his extraordinary geometrical imagination in solving this intellectual puzzle. The moon has two hemispheres, Subvolva and Privolva, divided not along the equator but along a circle through the poles. The revolving earth, or Volva, remains stationary in the Subvolvian sky, while the inhabitants of Privolva never get to see Volva. Seasonal changes are not as severe on the moon as on the earth because the moon's axis is tilted only 5 degrees according to Kepler (6½ degrees by modern reckoning) compared with the earth's inclination of 23½ degrees.

Somnium ends as Kepler awakens from his dream, leaving

behind Fiolxhilde, her son Duracotus, and the demon narrator, and finds himself in his own bed.

The influence of Kepler's *Somnium* on English literature goes beyond science fiction. Samuel Butler satirizes popular interest in the moon in his poem, "The Elephant in the Moon." Direct references to *Somnium* are apparent.

> Quote he—Th' Inhabitants of the Moon,
> Who when the Sun shines hot at Noon,
> Do live in Cellars underground
> Of eight miles deep and eighty around
> (In which at once they fortify
> Against the Sun and th' Enemy)
> Because their people's civiler
> Than those rude Peasants, that are found
> To live upon the upper Ground,
> Call'd Privolvans, with whom they are
> Perpetually at open War.

Marjorie Hope Nicolson has suggested that there are echoes of *Somnium* in Milton's *Paradise Lost*. Satan's flight through Chaos is clearly reminiscent of the type of cosmic voyage described in *Somnium*. Satan temporarily changes to a form that recalls Kepler's lunar creatures, as he

> O'er bog, or steep, through strait, rough, dense, or rare.
> With head, hands, wings, or feet, pursues his way,
> And sinks, or swims, or wades, or creeps, or flies.
>
> (II, 948–950)

The third Hell described in Book II of *Paradise Lost* recalls the mountains, caves, and fissures of the lunar landscape. Finally, the indirect influence of *Somnium* can be seen in the moon voyages of Jules Verne and H.G. Wells.

Jonathan Swift

A little less than a century after Kepler's death, Jonathan Swift published *Gulliver's Travels* (1726). Meanwhile Isaac Newton had published his *Principia* and had laid the foundations of a new scientific age, as will be discussed in Chapter 4. The Newtonian picture of the universe had little need for God except as a remote First Cause which had long ago set the world-machine in motion. God's supreme gift, man's intellect, was sufficient to discover the smallest details of the workings of the machine. The mystery, magic and fate of Kepler's age had receded before the blazing light of reason and the inquiring mind.

Gulliver's Travels qualifies as an ancestor of modern science fiction because the author went to extraordinary lengths to surround the events of his satire with a plausible factual milieu. The details of each society that Gulliver visits are consistent with the basic assumptions and physical limitations of the members of that society. This verisimilitude makes Swift's trenchant observations all the more deadly, and separates *Gulliver's Travels* from the realm of pure fantasy literature. Criticism of existing society, veiled or otherwise, is another element which *Gulliver's Travels* has in common with much of modern science fiction.

Scientists and mathematicians are the objects of Swift's barbs in Part III, "A Voyage to Laputa." The name of the flying island of Laputa probably comes from the Spanish *la puta*, "the whore." According to Swift, scientists readily sell their talents to any patron with sufficient money. Swift's criticism applies to royal patronage as well. By accepting the charter (1662) from King Charles II, the Royal Society of London also accepted certain "strings"; Roman Catholics could not be members, for example.

An inhabitant of Laputa perfectly fits the stereotype of the absent-minded professor. Whenever they are supposed to speak or listen, citizens have to be gently hit by an attendant with an inflated bladder or "flapper." Their two obsessions are mathematics and music, the former finding expression in their description of a beautiful woman in terms of rhombs, circles, parallelograms and the like. Even their food is concocted in the form of geometrical diagrams or musical instruments. Naturally their clothes are ill-fitting and their houses in disrepair, because they devote all of their time to pure science.

> . . . this defect ariseth from the contempt they bear to practical geometry, which they despise as vulgar and mechanic, those instructions they give being too refined for the intellectuals of their workmen, which occasions perpetual mistakes. (Chap. II)

One would expect such superfluity of science would dispel fear and superstition, but Laputans are even more fearful than most. They have three fears connected with the sun. First, that the earth would fall into the sun. Newton had demonstrated in his *Principia* that the earth is continually falling into the sun, but that this motion is continually counteracted by the earth's tangential velocity which is at right angles to the fall. The balance between these two motions might be altered if the earth were impeded—by a dense cloud of dust, for example—but Newton had reckoned that this was extremely unlikely. Second, they feared that the sun would be encrusted with its own effluvia and cease to shine. Sunspots were cited as early warnings of this process. Third, they worried that the sun might use up all of its fuel.

In addition to these fears with regard to the sun, the Laputans were terrified by the reappearance of a comet whose tail might brush the earth with disastrous consequences. Swift says that the return of the comet is to be in "one and thirty years," referring to the famous Halley's comet which appeared in 1607 and 75 years later in 1682. It was expected back in another 75½ years or 1757 (31 years from the date of first publication of *Gulliver's Travels*). Halley himself had predicted the return in 1758, based upon his calculations of perturbations of the orbit of the comet by planets. (It actually reached perihelion in 1759, delayed by perturbations of Jupiter and Saturn.) These finer details of the prediction did not make nearly as big an impression upon the public as the paper Halley read before the Royal Society which explained the Biblical story of Noah and the flood by invoking a close encounter between the earth and a comet. According to Halley such an event could change the position of the poles and rotation of the earth, and cause great flooding.

In Chapter III Swift mentions that the Laputans have discovered two satellites of Mars, one three planetary diameters away, the other five diameters away. Swift probably based his allusion to the two Martian satellites on a speculation of Kepler's: since Venus has no moons, Earth has one, and Jupiter has four (after Kepler's lifetime, more were discovered), Mars ought to have an intermediate number befitting its position. He guessed two rather than three. Swift also gives the periods of these satellites (10 hours for the inner one, 21.5 hours for the outer); the distances and periods conform to the predictions of Kepler's Harmonic Law. About 150 years after *Gulliver's Travels* was published, two satellites of Mars, Phobos and Deimos, were actually discovered. Their respective distances and periods (1.4 and 3.5 diameters of the planet, 7.7 and 30.2 hours) are quite different from Swift's speculation.

The Grand Academy of Lagodo (the capital city) may be taken as a satire of the Royal Society. The members of the Academy are called "projectors" and their schemes "projects," a term which in contemporary usage referred to get-rich-quick schemes of a highly speculative nature. The terms that members of the Royal Society preferred to use of themselves and their work were "virtuosi" and "experiments."

When laymen try to interpret the titles and contents of scientific articles, science can easily become an object of ridicule. Witness Senator William Proxmire's "Golden Fleece Award" to the biggest, most ridiculous or most ironic example of government spending or waste. Some "fishy" research meriting the Golden Fleece is a study—the effects of alcohol on aggressive behavior in a species of sunfish—sponsored by the National Institute on Alcohol Abuse and Alcoholism. Senator Proxmire suggests that NIAAA may be interested in testing what it means to be "stewed to the gills" or to "drink like a fish"! Swift exploits the same technique to hoist science by its own pitard. Marjorie

Hope Nicolson has pointed out that Swift made use of actual papers presented to the Royal Society and published in *Philosophical Transactions of the Royal Society*. An example of this technique is the projector who is trying to make silk stockings from spider's webs, an actual suggestion of a Frenchman, M. Bon, printed in the *Philosophical Transactions* of 1710. In another paper in 1708 a Dr. Wall praises certain animal-produced colors which are more permanent than artificial dyes. Swift has imaginatively combined the two papers into one of the projects of the Academy.

The most famous project is that of the projector who

> had been at work eight years upon a project for extracting sunbeams out of cucumbers, which were to be put into vials hermetically sealed, and let out to warm the air in raw inclement summers. (Chap. V)

Probably this account was loosely based upon the reports of Stephen Hales to the Royal Society on plant and animal physiology. The process of photosynthesis was unknown, but it had been observed that sunlight was needed for plant growth, and that large quantities of air were released by some plants, notably apples. Swift's projectors reverse this process to obtain sunlight from plants. If Swift did have Hales' reports in mind when he wrote the above passage, he greatly underestimated their significance. A plate in Hales' *Vegetable Statics* (1727) shows his famous invention, the pneumatic trough. The apparatus facilitates the collection of gas by separating the source of gas (an iron retort) from the receiving vessel ("the inverted chymical receiver") by a bent tube of lead. Using the pneumatic trough, Antoine Lavoisier identified the role oxygen played in combustion, and began a revolution in chemistry.

Though most of Swift's criticisms hit their target, the Royal Society was slow to make an institutional reply. The immediate occasion for the response was John Hill's criticism of many papers in the *Philosophical Transactions*. In the 47th volume of the *Philosophical Transactions* (1753), an Advertisement was inserted to the effect that now a committee of members would select papers for publication; previously, the printing of the *Transactions* was "the single act of the respective secretaries." The committee would not "answer for the certainty of their facts, or propriety of the reasonings contained in the several papers so published, which must still rest on the credit or judgment of their several authors." The Advertisement pointed out that the thanks proposed from the chair were a matter of civility to the authors of papers read at the meetings.

> The like is also to be said with regard to the several projects, inventions, and curiosities of various kinds, which are often exhibited to the Society; the authors whereof, or those who exhibit them, frequently

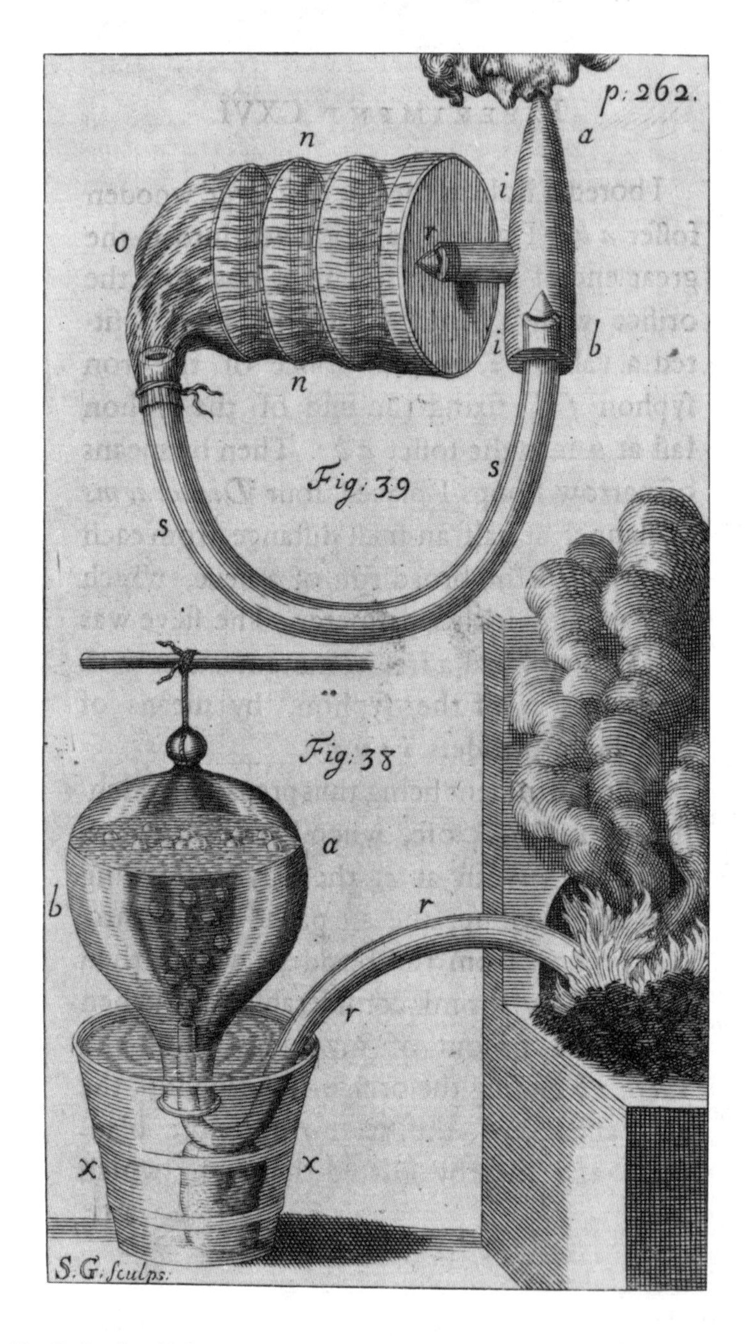

Fig. 7. Stephen Hales' pneumatic trough is shown in the lower half of the figure. From Vegetable Statics *(1727).*

> take the liberty to report, and even to certify in the public newspapers, that they have met with the highest applause and approbation. And therefore it is hoped that no regard will hereafter be paid to such reports and public notices; which, in some instances, have been too lightly credited, to the dishonour of the Society.

After this reform in selection of papers, the average quality of the papers printed greatly improved.

Madness overtook Swift at the end of his life, making him "a driv'ler and a show" in Samuel Johnson's words. Hints of Swift's impending breakdown surface at the end of *Gulliver's Travels.* When Gulliver returns home he cannot endure the sight and smell of his own family, reflecting a misanthropy in Swift that is surely abnormal. Swift lost his reason at a time when larger transformations were taking place in his society. As the eighteenth century drew to a close, the Age of Reason was giving way to a glorification of unreason, a revival of interest in the occult, a rediscovery of the mysteries of the unconscious and Nature; in short, the dawn of the Romantic Age.

Mary Shelley

It has never ceased to amaze succeeding generations that the novel *Frankenstein* was written by a girl of nineteen. Not just *any* girl wrote this, but Mary Wollstonecraft Godwin, the daughter of two well-known authors, and the lover and later wife of Percy Bysshe Shelley. The occasion for the writing of the novel was a house party in Geneva where the Shelleys and Mary's stepsister were neighbors of Lord Byron and his physician, John William Polidori. At Byron's suggestion each member of the party wrote a ghost story. Besides Mary's story, only one other was ever published, a fragment of Byron's about vampires, later completed by Polidori, and probably used by Stoker as one of the sources for *Dracula.* Mary published her book anonymously in 1818, adding her name to a subsequent edition in 1831. She wrote in the tradition of the "Gothic" novel, a style which began with Horace Walpole's *The Castle of Otranto* (1765) and still flourishes today. However, Mary Shelley eschewed the lurid paraphernalia of the Gothic romance, as she says in the Preface to *Frankenstein,*

> I have not considered myself as merely weaving a series of supernatural terrors. The event on which the interest of the story depends is exempt from the disadvantages of a mere tale of spectres or enchantment. . . . I have thus endeavoured to preserve the truth of the elementary principles of human nature, while I have not scrupled to innovate upon their combinations.

This realism singles out *Frankenstein* among the many Gothic novels as an ancestor of modern science fiction.

Mary Shelley does not describe the means by which Victor

Frankenstein brings life to the lumps of charnel-house flesh he has assembled, but she hints that the agency is electrical. Victor says at one point,

> Before this I was not unacquainted with the more obvious laws of electricity. On this occasion a man of great research . . . entered on the explanation of a theory which he had formed on the subject of electricity and galvanism, which was at once new and astonishing to me.
> (Chap. II)

The reference is to Professor Galvani of Bologna who discovered in 1780 that a pair of frog legs would contract whenever there was a spark. To the readers of the accounts of his experiment a logical next step was to assume that electricity might animate dead frogs. The author gives another suggestion of electrical animation when Victor speaks of trying to "infuse a spark of being" into his progeny.

The classic Hollywood film of *Frankenstein* (1931) directed by James Whale featured Boris Karloff as the monster (or "android" in the jargon of modern science fiction). Its great success led Universal Studios to produce a seemingly endless series of sequels: *Bride of Frankenstein* (1935), *Son of Frankenstein* (1939), *Ghost of Frankenstein* (1942), *Frankenstein Meets the Wolf Man* (1943). After the Second World War an element of parody dominates the series: *Abbott and Costello Meet Frankenstein* (1948), *I Was A Teenage Frankenstein* (1957), and *Frankenstein and the Space Monsters* (1965). The only element of the original story retained in these later pictures was the name, unfortunately not copyrighted. The monster appears in the television comedy series, *The Munsters*. Recently Christopher Isherwood and Don Bachardy produced a script for a television version, "Frankenstein: The True Story," which has many similarities with the original. here the monster is portrayed as a beautiful young man whose features gradually decay, leading him to hate himself and blame his creator. Mel Brooks gives a satiric treatment in his production of *Young Frankenstein* (1974). Shot in black and white with a real fondness for the horror films of the thirties, this film has the novelty that both monster and maker live happily ever after.

The British film industry was not far behind the American with Hammer Studios color productions of *The Curse of Frankenstein* (1957), *Frankenstein Created Woman* (1967), *Frankenstein Must Be Destroyed* (1969), and *The Horror of Frankenstein* (1970). The above lists are incomplete, but they give an idea of the public's continuing interest in this theme.

Realistic details alone do not account for the great popularity of Mary Shelley's story. She tapped, knowingly or otherwise, deep springs of myth. The subtitle of the novel is "The Modern Prometheus," a reference to the Greek legend related by Hesiod. In

Fig. 8. Boris Karloff as Frankenstein's monster from the original 1931 film. (Photo courtesy Museum of Modern Art Stills Archive).

the legend Zeus hid fire from the human race after being tricked by Prometheus, who subsequently stole the fire from heaven for the benefit of earth. Zeus punished him by chaining him to a rock and having an eagle feed daily upon his liver until Hercules freed him. Aeschylus, in his play *Prometheus Bound,* expands the role of Prometheus from giver of fire to giver of intelligence, mathematics, writing, domesticated animals, and the healing arts. Percy Shelley further develops the legend in *Prometheus Unbound,* where Prometheus becomes in Shelley's words, "the type of the highest perfection of moral and intellectual nature, impelled by the purest and truest motive to the best and noblest ends." Beethoven uses the legend for the ballet music, "The Creatures of Prometheus." In the novel the author gives the role of Prometheus to Victor Frankenstein who has all the benefits of loving family and excellent education. To "examine the causes of life" is his great quest, but in so doing he usurps the powers of the Creator, the equivalent of stealing fire from heaven.

Several other myths dealing with inanimate forms being brought to life also have a bearing on the Frankenstein story. The Roman poet Ovid relates the story of a sculptor, Pygmalion, who falls in love with his own statue of the ideal woman. The goddess Venus takes pity upon him and brings the statue to life. In the Bernard Shaw play of the same name, Pygmalion becomes Henry Higgins, teacher of phonetics, who falls in love with a cockney flower girl. This in turn was made into the successful stage and screen musical "My Fair Lady." From the Middle Ages comes the Jewish legend of the Golem, an effigy which could be animated by placing a charm in its mouth or on its head, the source of the German silent film classic, *Der Golem* (1920). More recently a poem by Goethe inspired Paul Dukas' tone poem "The Sorcerer's Aprentice," music for a segment of the Walt Disney film, *Fantasia,* in which a broomstick is animated, becoming at first an obedient servant, but ultimately uncontrollable. The android created by Victor Frankenstein has the distinction of being the only one up to that time brought to life by scientific means.

Besides embodying mythic traditions, the novel also possesses a psychological validity. Before Sigmund Freud's revolutionary publications in psychoanalysis, no adequate language existed to express the reality of the unconscious. Consequently, many authors used symbols or dramatized these kinds of insights as Mary Shelley did in *Frankenstein*. According to this psychological viewpoint, the monster represents another aspect of Victor Frankenstein's personality, the part which craves companionship and love, things which the conscious Frankenstein had denied. Unfortunately the monster also has Frankenstein's repressed destructive impulses. Frankenstein's "case" might be diagnosed today as "narcissistic schizophrenia."

Mary Shelley adroitly manipulates the three-layered plot of *Frankenstein*. First she introduces Robert Walton, a gentleman turned Arctic explorer. He rescues Victor Frankenstein from an ice floe upon which he had been trapped while pursuing the monster. Frankenstein recognizes in Walton a kindred spirit, and urges him to take his ship farther north so that Frankenstein may find the monster and destroy him. In the second layer, the main bulk of the novel, Victor Frankenstein tells how he created the monster who subsequently ran amok. Frankenstein intended his creation to be beautiful, but it didn't quite turn out that way.

> His limbs were in proportion, and I had selected his features as beautiful. Beautiful!—Great God! His yellow skin scarcely covered the work of muscles and arteries beneath; his hair was of a lustrous black, and flowing; his teeth of a pearly whiteness; but these luxuriances only formed a more horrid contrast with his watery eyes, that seemed almost the same color as the dun white sockets in which they were set, his shrivelled complexion and straight black lips. (Chap. V)

A subsequent meeting of Frankenstein and monster at the top of a mountain affords an opportunity for the novelist to give a third version of the story from the point of view of the monster.

The most serious deficiency of the plot is the wildly improbable way the monster achieves his education after being abandoned by his creator. The monster finds shelter in a vacant hovel attached to the cottage of a peasant family. Through a providential chink in the wall he observes the devotion of a blind father to his son and daughter. When the son's Arabian girl friend visits the cottage, the son gives her English lessons, affording the monster convenient auditing privileges. The monster realizes that he has been missing companionship and love, and yearns for a mate. He promises to live in the jungles of South America if Frankenstein will consent to produce a female monster. Frankenstein reluctantly agrees and retires to the Orkney Islands for this purpose. Frankenstein changes his mind midway through his project in order to spare the world the horrible progeny that the union of two monsters might produce. Frankenstein's refusal to create a mate for the monster arouses him to further villainy (he had already murdered Frankenstein's brother William and framed the family servant for the crime). The final blow is the monster's murder of Frankenstein's bride Elizabeth on her wedding night. Frankenstein then pursues the monster with intent to kill, while the monster lures Frankenstein farther and farther north in order to expose him to the agonies of the Arctic winter.

The role reversal taking place between Frankenstein and his creation is fascinating. This reversal, in which slave becomes master and *vice versa*, has been so complete that the general public identifies the name Frankenstein with the monster instead of his creator. In fact the monster does not have a name, reinforcing the idea that the processes of socialization have been unavailable to this lonely creature. The reader's sympathies are drawn toward the monster who says poignantly that,

> no father had watched my infant days, no mother had blessed me with smiles and caresses; or if they had, all my past life was now a blot, a blind vacancy in which I distinguished nothing. From my earliest remembrance I had been as I then was in height and proportion. I had never yet seen a being resembling me, or who claimed intercourse with me. What was I? (Chap. XIII)

Indeed Frankenstein abandoned his creation from the moment he first opened his baleful eye. The monster reminds Frankenstein of his responsibilities:

> Oh, Frankenstein, be not equitable to every other, and trample upon me alone, to whom thy justice, and even thy clemency and affection, is most due. Remember, that I am thy creature; I ought to be thy Adam; but I am rather the fallen angel, whom thou drivest from joy for no misdeed. Everywhere I see bliss, from which I alone am irrevocably excluded. I was benevolent and good; misery made me a fiend. Make me happy, and I shall again be virtuous. (Chap. X)

The monster is not totally fiendish, since he experiences remorse at the end of the novel. In the final image, the monster bends over the dead body of Frankenstein and asks his forgiveness.

Mary Shelley was the first to introduce another theme which frequently appears in subsequent science fiction. In her novel *The Last Man* she explores at great length the feelings of Lionel Verney, the single remaining individual after plague has destroyed the entire population of earth in the twenty-first century. Unfortunately, the author only fitfully realizes the dramatic possibilities inherent in this situation.

Nathaniel Hawthorne

Although the American Nathaniel Hawthorne lived during the reign of Queen Victoria, he was more of a romantic than a Victorian. His writings contain reveries, allegories, strange transformations "seen through a glass darkly," to use a Biblical phrase. Several of his works are science fiction stories according to the definition discussed at the beginning of this chapter. They have certain themes in common. The scientist is either depicted as a loner as Dr. Heidegger in "Dr. Heidegger's Experiment" (1837) and Owen Warland in "The Artist of the Beautiful" (1844), or at most assisted by a single individual. The assistant to the scientist Aylmer in "The Birthmark" (1843) is Aminadab, who rejoices when his master's experiment goes awry. As Aylmer's wife dies, "he heard a gross, hoarse chuckle, which he had long known as his servant Aminadab's expression of delight." Beatrice in "Rappaccini's Daughter" (1844) is the prototype of the mad scientist's daughter, torn between her love for her father and her suspicions concerning his experiments. The scientist reacts with hostility when intruders disturb his experiments. Here is Aylmer's reaction when his wife Georgina visits his laboratory for the first time:

> Aylmer raised his eyes hastily, and at first reddened, then grew paler than ever, on beholding Georgina. He rushed toward her and seized her arm with a grip that left the print of his fingers upon it.
>
> 'Why do you come hither? Have you no trust in your husband?' cried he impetuously. 'Would you throw the blight of that fatal birthmark over my labors? It is not well done. Go, prying woman, go!'

The outburst is all the more remarkable considering that his wife is the subject of the experiment, and should have the right to know the ingredients of the potion and the risk she is taking. When Dr. Heidegger's chambermaid lifts one of his books, we get this amusing picture.

> The skeleton had rattled in its closet, the picture of the young lady had stepped one foot upon the floor, and several ghastly faces had peeped forth from the mirror; while the brazen head of Hippocrates frowned, and said,—'Forbear!'

Hawthorne didn't single-handedly invent the mad scientist. I have discussed Swift's previous use of the balmy, but harmless, projectors in satirizing the Royal Society. Mary Shelley's portrait of Victor Frankenstein includes a strong element of gothic horror. Hawthorne takes the image of the mad scientist a step further. His scientists—Heidegger, Aylmer, Rappaccini—lack the innocence of the philosophers of Laputa. They are both utopian and Faustian. To a certain degree the literary portrayals of scientists as anti-social individuals without natural affections reflected popular attitudes of the time, so the authors should not bear all of the blame for creating these stereotypes. Hawthorne writes with a great deal of sympathy for his solitary characters because he himself preferred to keep his own company. When he was fifty-four he said, "I doubt whether I have ever really talked with half a dozen persons in my life, men or women."

Subsequent authors have frequently employed another motif appearing in the Hawthorne stories, the use of drugs. Sometimes his characters take drugs to achieve a desired goal: Dr. Heidegger's friends grasp at youthfulness and gaiety, and Georgina yearns for the removal of her birthmark. More ominously, Rappaccini uses a drug on his daughter without her consent. In each case the effects are not promising; Dr. Heidegger's elixir produces only temporary effects, and the other drugs result in death. Note that none of the drug producers partakes of his own cordial.

Rarely do they achieve the disinterestedness associated with scientists; what they feel is more like the attitude of a cat watching a mouse. Dr. Heidegger demurs with the following words,

> For my own part, having had much trouble in growing old, I am in no hurry to grow young again. With your permission, therefore, I will merely watch the progress of the experiment.

Of the four stories mentioned above, I think that "The Birthmark" is the most profound, and deserves further scrutiny. The story is a parable complete with moral at the end. Georgina is the most beautiful woman on earth, a "specimen of ideal loveliness," save for one flaw, a crimson birthmark on her cheek in the shape of a tiny hand. This "sole token of human imperfection" may be a reference to her sensuality. The scientist Aylmer marries her and becomes obsessed with the birthmark. He devotes all of his efforts to devise ways of removing it. Hawthorne describes Aylmer's assistant Aminadab as gross and earthy, contrasting with Aylmer's air of spirituality. Georgina, who represents the non-scientific world, acquiesces in her husband's plans. She says,

> There is but one danger—that this horrible stigma shall be left upon my cheek! Remove it, remove it, whatever be the cost, or we shall both go mad.

The cost is indeed high, since Georgina dies just as the final potion takes effect and the birthmark fades away. Is Aylmer an evil monster for killing his wife? In some ways he seems a confused and tragic hero. His sincerity in wanting to improve upon nature is obvious, but it has tragic consequences because the imperfection he wants to remove is part of human nature. His mistake seems to be an identification of science with religion. The language of the story is filled with religious words: "spiritual affinity"; Aylmer says, "Ah, wait for this one success, then worship me if you will"; and the liquid from the goblet (the word itself has religious overtones) is "like water from a heavenly fountain."

What are the dangers that Hawthorne might have seen in science-as-religion at the time the story was written? His specific scientific allusions appear harmless enough. To sooth and amuse Georgina, Aylmer shows her some optical demonstrations and takes her picture. Hawthorne may be referring to the diorama (1822) and daguerrotype (1835) invented by Daguerre and the stereoscope (1832) invented by Wheatstone. These devices were intended as entertainments, scientific playthings, and were not the sort of scientific achievements to induce worship on the part of anyone. To see what Hawthorne was warning against we must examine the state of scientific knowledge in the first half of the nineteenth century.

The public couldn't distinguish between science and pseudoscience—physiognomy, phrenology, homeopathy, and mesmerism. How could people accurately evaluate the pseudosciences' exaggerated claims to affect health? Each pseudoscience contained enough of a kernel of truth to keep the myth of its efficacy alive. Hawthorne's wife Sophia received beneficial treatment for headaches from a mesmerist. But Edgar Allen Poe capitalized on the animal-magnetism craze in his story, "The Facts in the Case of M. Valdemar." Poe's fabrication, in which a dying man is magnetized just before the moment of death and held in suspended animation, was widely accepted. Hawthorne knew public taste and aimed to appeal to it. Seen in this light, Hawthorne was speculating on what might transpire if the possibilities implied by pseudoscience become reality.

Meanwhile, several developments in nineteenth-century science gave rationalists hope that life itself could be explained on a mechanistic basis, a scheme so obviously successful in the Newtonian picture of the heavens. C. Lyell announced his uniformitarian theory in his book *Principles of Geology* (1830-33). According to Lyell, natural forces still acting today produced the phenomena of geology, rather than catastrophes such as a universal flood. He accepted fixity of species, and deduced that new fauna had been created in every age, only to become extinct. Thus it became increasingly difficult to believe in the biblical story of creation. The achromatic miscroscopes of G. Amici (1827) made possible new insights into the structure of tissues. Charles Darwin had taken his voyage aboard the Beagle (1831–1836), but had not yet published his *Origin of Species.* M. Schleider and T. Schwann propounded a cell theory describing the body as a colony of cells originating from the union of two cells, sperm and egg. Chemists were on the verge of discovering the first artificial aniline dye, magenta. Thus Hawthorne was writing at a time when science had taken great strides, and was on the threshold of making breakthroughs in physiology, pathology, histology, bacteriology, and organic chemistry. His premonitions regarding the dangers of experiments on humans were therefore timely.

This chapter has summarized two centuries of early science fiction. The various writers have explored strikingly different themes. Kepler speculated on the problems of space travel while laying the foundation for our modern understanding of planetary orbits. Swift criticized scientific discoveries as useless and trivial; nevertheless, he acknowledged man's growing power over nature. In the nineteenth century, scientists began to understand some of the incredibly complex life processes. Mary Shelley and Hawthorne foresaw that science was rapidly approaching a stage where it could alter and manipulate human beings. Amis' definition of science fiction at the beginning of the chapter referred to the innovations of science and technology. Science fiction writers and the public were hardly prepared at the mid-point of the nineteenth century for the accelerating rate at which these innovations were coming.

Bibliography

Aldiss, Brian W., *Billion Year Spree* (Doubleday, 1973).

Amis, Kingsley, *New Maps of Hell* (V. Gollancz, Ltd., London, 1961).

Anobile, Richard J., Ed., *Frankenstein* (Universe Books, New York, 1974).

Ash, Brian, *Faces of the Future* (Taplinger Pub. Co., New York, 1975).

Bailey, J.O., *Pilgrims Through Space and Time* (Argus Books, New York, 1947).

Barron, Neil, *Anatomy of Wonder* (R.R. Bowker, New York, 1976).

Beer, Arthur, and Beer, Peter, eds., *Kepler: Four Hundred Years*, Vistas in Astronomy, Vol. **18** (Pergamon Press, Oxford, 1975).

Bretnor, Reginald, *Science Fiction, Today and Tomorrow* (Harper, New York, 1974).

Caspar, Max, *Kepler* (Abelard-Schuman, London, 1959).

Florescu, Radu, *In Search of Frankenstein* (New York Graphic Society, Boston, 1975).

Franklin, H. Bruce, *Future Perfect: American Science Fiction of the Nineteenth Century* (Oxford Univ. Press, New York, 1966).

Green, R.L., *Into Other Worlds: Space Flight in Fiction from Lucian to Lewis* (Abelard-Schuman, London, 1958).

Holton, Gerald, "Johannes Kepler's Universe: Its Physics and Metaphysics," Am. J. Phys. **24,** 341, 1956.

Kepler, Johann, *Somnium,* Rosen, Edward, tr. (Univ. of Wisconsin Press, Madison, 1967).

Koestler, Arthur, *The Watershed* (Anchor Books, Garden City, 1960). A section of *The Sleepwalkers.*

Moers, Ellen, "Female Gothic: The Monster's Mother," N.Y. Rev. of Books, **21,** 24, March 21, 1974.

Moore, Patrick, *Science and Fiction* (George G. Harrap, London, 1957).

Moskowitz, Samuel, *Explorers of the Infinite* (World, 1963).

Nicolson, Marjorie Hope, *Science and Imagination* (Cornell Univ. Press, Ithaca, 1956).

____. *Voyages to the Moon* (Macmillan Co., New York, 1948).

Pauli, W., *The Influence of Archetypal Ideas on the Scientific Theories of Kepler* in *The Interpretation of Nature and the Psyche* (Pantheon Books, Bollingen Series, LI, New York, 1955).

Postl, Anton, "Kepler's Anniversary," Am. J. Phys. **40,** 660, 1972.
Shelley, Mary W., *Frankenstein* (Oxford Univ. Press, 1969).
____. *The Last Man* (Univ. of Nebraska Press, Lincoln, 1955).
Small, Christopher, *Mary Shelley's Frankenstein* (Univ. of Pittsburgh Press, 1973).
Tropp, Martin, *Mary Shelley's Monster* (Houghton Mifflin Co., Boston, 1976).

Chapter 3

Scientists On The Stage In The Twentieth Century

IN THE 20TH CENTURY, science moved beyond the capability of altering and manipulating human life to the possibility of creating a humanoid machine. Two plays which probed the dangers of machines replacing humans appeared in the period between the two world wars: *R.U.R. (Rossum's Universal Robots)* (1923) by Karel Čapek, and *The Adding Machine* (1923) by Elmer Rice. Both plays have since become classics of the theater, but a more immediate danger pushed the problems of automation into the background. The seeds of a nuclear holocaust had been sown in Switzerland in 1905 when Albert Einstein spelled out the mass-energy equivalence in the seemingly innocuous formula, $E=mc^2$. In the thirties in Berlin, O. Hahn and F. Strassman discovered the fission of uranium, making possible, at least in theory, a chain-reaction weapon. Hitler's embargo on the export of uranium from Czechoslovakia spurred an American effort to beat the Nazis to the atomic bomb.

American research in nuclear science during the interwar period began as a low-budget affair. To give an example, the first cyclotron which Ernest O. Lawrence constructed (1930) was a small, pitiful-looking contraption held together with red sealing wax. But the days of Big Science were not far off. Lawrence was soon able to interest philanthropists in donating to his projects because cyclotron beams had the potential of destroying cancer cells in patients. With the advent of the Second World War, nuclear physics took on supreme importance in the struggle to develop the atomic bomb. Lawrence proposed an electromagnetic method of separating the uranium isotopes, a scheme based upon his experience with large magnets and vacuum systems in the cyclotron work. Government money was made available, and the Lawrence-inspired devices (calutrons) were built. In the end, they were less successful than the gaseous-diffusion process established at Oak Ridge. The money spent on Lawrence's project was tiny compared to the total outlay. The physicists soon realized that they had made a Faustian bargain, and some of them demurred at the strings attached. They found the secrecy imposed on their work oppressive, and many hoped that the atomic bomb that they were developing would never have to be used.

Among the many scientists contributing to the American atomic bomb effort, playwrights have singled out Albert Einstein and J. Robert Oppenheimer. The life of Oppenheimer has some striking parallels to Galileo Galilei's life, so I will first examine the scientific careers of these two men and the plays written about them. After considering Einstein's contributions to science, I will discuss a play and an opera written about him. I will conclude by examining another opera based on the life of a famous scientist, Johannes Kepler, whose scientific accomplishments have been detailed in the previous chapter.

Galileo Galilei

A Scientific Profile. Shakespeare and Galileo were born the same year (1564); Shakespeare's literary works are similar to Galileo's scientific works in their originality, profusion, and subsequent influence.

Galileo's telescopic discoveries are the most widely known. He designed and constructed a telescope after hearing reports about the instrument from the Netherlands. Merely knowing the existence of a working device often facilitates construction of another one. (The Russian atomic bomb effort had this advantage plus the reports of spies such as Klaus Fuchs). Although Galileo did not invent the telescope, he was apparently the first to direct it to the heavens and correctly interpret what he saw. To him belongs the credit for identifying mountains on the moon, for seeing the rings of Saturn (albeit rather indistinctly) and phases of Venus, and for realizing that the Milky Way was composed of myriads of individual stars. He not only discovered four satellites of Jupiter, but worked out tables which gave their periods of revolution and times of eclipse behind the planet with the purpose of using these data for determining longitude at sea. Unfortunately, the scheme was impractical because it proved impossible to sight the satellites from the tossing deck of a ship. He was not the first to observe sunspots, nor did he identify them properly (he thought they were clouds above the sun's surface), but he did deduce that they were evidence for the sun's rotation.

In the book which got him into trouble, *Dialogues Concerning the Two Chief World Systems,* Galileo brilliantly manipulates the dialogue form, following Plato's example. By a series of adroitly handled questions and responses, Galileo could set up his adversaries and have his characters say with wit and irony what he

Fig. 9. An engraving of Galileo from the 1744 edition of Dialogues Concerning the Two Chief World Systems.

himself could not say openly concerning the Copernican heliocentric hypothesis. Thus, his heterodox opinions might have been tolerated by the authorities had not Galileo put the Pope's unanswerable logic in the mouth of a straw-man, Simplicio, at the end of the book. Unfortunately for Galileo, Pope Urban VIII lacked a sense of humor.

When Galileo was first ordered by the Inquisition (1632) to stand trial at Rome, he was relatively confident that he held the trump card, a statement in Cardinal Bellarmine's handwriting of seventeen years before to the effect that, although the doctrines of Copernicus could not be defended or held as fact, they could nevertheless be advanced as hypotheses. Incredibly, the Inquisition had another piece of paper stating that in addition Galileo had been enjoined not to hold, teach, or defend the Copernican doctrines in *any way whatsoever, verbally or in writing*. The original Bellarmine certificate still exists in Vatican files, as does the Inquisition version. The consensus of modern historians is that the Inquisition minute is a false document because it lacks both Galileo's and Bellarmine's signatures, is not notarized, and is witnessed only by members of Bellarmine's staff. Furthermore, it is not dated, and begins on the back of another document, hardly the legal cornerstone upon which to build a case.

Galileo was examined three times by the Inquisition, and he was sentenced on June 22, 1633 to abjure, curse, and detest his errors and heresies—that the sun is the center of the world and immovable, and that the earth is not the center of the world and moves. In addition, the *Dialogues* was prohibited by public edict. For penance he had to recite the seven penitential Psalms once a week for three years, and, more damaging to his person and his career, he was sentenced to prison. The illegality of these proceedings, based as they were on the questionable Bellarmine document, evidently disturbed some of the Cardinals to the extent that three of the ten did not sign the sentence. One of the abstainers was the Pope's nephew, Francesco Barberini.

Galileo went through the motions of formal recantation in the Domenican Convent of Santa Maria sopra Minerva. After the prison sentence was commuted to house arrest, he retired to his farm in Arcetri where he lived out the remaining eight years of his life, fighting against his oncoming blindness. But he was not as effectively silenced as the Inquisition had planned.

After the trial and recantation, when it seemed that his public career was over, he managed to smuggle out to the Netherlands for publication of his greatest work, the *Discourses Concerning Two New Sciences*. (The two new sciences were strength of materials and kinematics.) In this book, the product of his mature years, he brilliantly combined mathematical analysis and experimentation. In fact, the modern concept of an experiment is due to him; he called it *cimento*[3] ("ordeal").

Among the most significant of the discoveries recorded in *Discourses* are his studies of pendulum motion, resonance, inertia, average speed and acceleration, and free fall in one and two dimensions. He showed that the ratio of the periods of two different pendulums was equal to the ratio of the square roots of their respective lengths. He was the first to notice the phenomenon of resonance: by sounding the string of one instrument, the corresponding string of an adjacent instrument could be set in motion. He introduced his concept of inertia (later generalized by Newton) in the following words: "Any velocity once imparted to a moving body will be rigidly maintained as long as external causes of acceleration or retardation are removed, a condition which is found only on horizontal planes." He also defined the concepts of *average* speed and acceleration. (Subsequently Newton and Leibniz gave a clear definition of *instantaneous* speed). He realized that rate of fall was independent of weight, but the famous story of dropping weights from the Leaning Tower of Pisa is most likely apocryphal. Since free fall due to gravity was too swift for the primitive time-keeping mechanisms he had at his disposal, he cleverly made use of an inclined plane to slow down the acceleration. The climax of his kinematical insights was his realization that, for a ball projected horizontally, the vertical fall due to gravity and the constant horizontal velocity (one a "natural", the other a "violent" motion according to Aristotle) proceeded independently. The resultant motion he identified as a parabola.

To twentieth-century readers the discoveries set out in the *Discourses* may seem simple, but they came as revelations to Galileo's contemporaries. His lucid exposition was like a fresh wind blowing away the accumulated clouds of centuries of anthropomorphic and mythical thinking. His successful blend of hypothesis, measurement, and mathematical analysis laid the foundation used by Newton and later Einstein as the basis for modern dynamics and kinematics. But he was not infallible. His arguments as to the causes of the tides were incorrect; the solution of that problem had to await a more complete theory of mechanics. In other areas he failed to avail himself of explanations more readily obtainable. As shown in the previous chapter, during Galileo's own lifetime Kepler had found the secret of unscrambling the complicated apparent motions of the planets. Curiously, Galileo failed to use Kepler's laws of planetary motion, which would have greatly strengthened his arguments for the Copernican hypothesis. This blind spot reveals a streak of unreceptivity, even arrogance, in his character, but without it he

[3]The name is preserved in the *Accademia del Cimento*, a society of experimentalists, of which Galileo's disciple Torricelli, the inventor of the barometer, was a member, and in *Nuovo Cimento* ("New Experimentation"), an Italian physics journal published to this day.

might not have fought as long as he did against the forces of unreason in his time.

The curtain has not yet rung down on the Galileo affair. A commission announced on October 23, 1980 by Bishop Paul Poupard, Vice-President of the Vatican's Secretariat for Nonbelievers, will review Galileo's case. Galileo may yet be acquitted, although posthumously.

On the Stage. At least three playwrights have used the swirl of events surrounding Galileo as the basis for biographical plays. Here, the twentieth-century stage works of Barry Stavis and Bertolt Brecht will be compared with an earlier version of Galileo's life by François Ponsard.

Barry Stavis' *Lamp at Midnight* (written in 1942) focuses on Galileo's telescopic discoveries. The play was more of a critical than a financial success; probably its seriousness and strict historicity discouraged immediate popular approval. Brooks Atkinson, drama critic of the *New York Times,* felt that the play was superior to Brecht's version of Galileo's life. According to Stavis, Galileo refuses to flee to England because he wishes to fight for his book and the truth. He realizes that Italians will not have the opportunity of reading his book unless it has the *imprimatur,* and he wants the truths it contains to penetrate the church itself. His will begins to yield when the Inquisitor impresses upon him that he is a man of chaos, destroying the unity of the church. Luther and Calvin attacked the church from without, Galileo is told, but he threatens the church from within. The Pope is worried; since Galileo's book is written in Italian and appeals to reason and not to faith, people may begin to reason independently, fragmenting the church into a thousand pieces. Finally, Stavis shows Galileo confounded by the Inquisition's version of Cardinal Bellarmin's injunction against promoting Copernican doctrine in any way whatsoever. Under such pressure, Galileo, as a loyal son of the church, signs the confession.

In contrast, Bertolt Brecht's play, *Galileo* (written in 1938, revised in 1947), freely distorts history for dramatic and propagandistic purposes. This may be seen in several instances. While in Padua, the real Galileo sired three children (Virginia, Livia, and Vicenzo) by a mistress, Marina Gamba. When he left Padua to become Chief Mathematician and Philosopher to Cosimo II, Grand Duke of Tuscany, he left Marina behind. Later, he arranged for both daughters to enter the convent of San Matteo in Arcetri outside Florence. Virginia became Sister Maria Celeste at age 16, making her lengthy engagement to Ludovico Marsili as in the play unlikely. This rewriting of history is necessary for Brecht to demonstrate how Galileo ran roughshod over his daughter's feelings, and how implacable was the opposition from the fami-

Fig. 10. Frontispiece from an early edition of Galileo's Dialogues Concerning the Two Chief World Systems, *showing a discussion among Aristotle, Ptolemy, and Copernicus; in the actual book the debate is carried on by three characters invented by Galileo.*

lies of the landed aristocracy, such as the Marsilis.

In another historical inaccuracy, Brecht's Marxist ideas are that safety lies in working for the rising bourgeoisie, and danger comes from an alliance of university professors, ecclesiastics, and the aristocracy. Actually, the Republic of Venice, with a show of characteristic independence, offered Galileo sanctuary via its intermediary and elder statesman Francesco Morosini, but Galileo felt that he had to obey his church.

Brecht seriously misrepresents seventeenth-century thinking when in Scene 5 he has the Old Cardinal go into hysterics because the earth is about to be displaced from the center of things. The Cardinal feels strongly that man is the object of God's attention on the central earth, and that the eye of the Creator is on man. However, just the opposite was characteristic of seventeenth-century thought. The place of honor was at the periphery, beyond the *Primum Mobile,* where God dwelled. At the very center was Hell. Man shared his habitation with the lower orders of life, at a consequent loss to his sense of importance and dignity. The earth was the scene of corruption and decay, but beyond the sphere of the moon were the incorruptible heavens.

Why does Brecht throw away the opportunity for dramatizing Galileo's trial? I think the drama is heightened by the playwright's device of reflecting the anguish of Galileo in the reactions of his family and friends. Ultimately, anguish turns to anger, and his friends and disciples feel betrayed by Galileo. The emphasis on the actions of groups of people has led some commentators to think that the hero of the play is not Galileo but the people. The masses do play a prominent role in two of the most important scenes in the play, Scene 9 and Scene 11. The Carnival Scene shows how Galileo's theories, much distorted and almost unrecognizable, are upsetting the usual social order; the cobbler's boy tries on the new pair of boots he is supposed to deliver, an egg is broken over a gentleman's head, etc. Obviously, the landed gentry and the church cannot tolerate this. The sound of shuffling feet is background to the Robing Scene, the transformation of Maffeo Barberini into Pope Urban VIII for a formal audience. At the beginning of the scene he resists the suggestions of the Inquisitor, but at the end of the scene, when only the face of Barberini remains in the midst of the splendid vestments, he gives in, and allows Galileo to be interrogated by the Inquisition. The shuffling of feet has reminded him of his official duties as Shepherd of the Flock. The masses are not heroes automatically, but require careful and patient education if they are to properly fulfill their role, as shown by the last scene. Andrea Sarti cannot convince a little boy that a woman is not a witch. The boy's preconceived notions are stronger than the evidence of his own eyes that she is stirring porridge, not riding a broomstick. At least the boy looked through the window of the woman's cottage; the Florentine professors had refused to go even that far when they declined to look through Galileo's telescope.

The popular understanding of the play is that it mainly concerns the conflict between science and the church. This is an oversimplification. A better statement would be that the play presents the scientists' tentative gropings toward truth and the church's absolute dogma as two different intellectual positions not easy to reconcile. Brecht scrutinizes both sides of this engagement. He blames Galileo for being too weak to champion the truth that he discovered. Nor does the church escape Brecht's censure. According to Brecht, capitalist power, not Christian conviction props up the church's dogma. So the church opposes Galileo because the capitalist system involving the entire social structure would collapse if Galileo's ideas took hold.

In Brecht's view, Galileo is a coward, because he capitulates upon being merely shown the instruments of torture. Brecht feels that Galileo always had the upper hand, and should have opposed the church more vigorously. Galileo's reaction is like Brecht's own when the House Un-American Activities Committee called him to testify and asked if he had ever made application to join the Comunist Party. Brecht answered negatively, and fled to East Germany. In the play, Galileo makes the speeches of a coward, but acts like a hero. The thirteenth scene contains a remarkable monologue for Galileo in which he explores the depths of compromise, cowardice, and capitulation. He speaks of his shameful surrendering of knowledge to the powers that be for them to abuse (this speech was revised after the American atomic bomb explosions). He feels that he has betrayed his profession, and should be drummed out of the corps of scientists. He surreptitiously produces the *Discourses* which he gives to Andrea Sarti for smuggling out of Italy. If he had resisted the power of the Inquisition, he would have been kept on much shorter rein, and couldn't have written at all. Since the *Discourses* rather than the *Dialogues* contain the basis for technological innovation and hence for the improvement of the lot of the masses, Brecht's Marxist interpretation is inconsistent when he chastizes Galileo for recanting.

A comparison of the above two plays with an earlier version of the life of Galileo reveals a nineteenth-century preoccupation with family values. *Galilée* (1867), written in rhymed verse by the French playwright François Ponsard, is not historically accurate, but Ponsard's distortions differ from those of Brecht. In this version Galileo is given a nagging wife who has much to say about science; she thinks Aristotle is right and Copernicus wrong. In the climactic scene of Act III, Galileo, his wife and daughter, her fiancé, and Galileo's student Vivian are all present in the Palace of the Inquisition in Rome, pleading with him to avoid a useless martyrdom. Galileo gives in essentially because his devotion to his family takes precedence over his duty to science, and asks God's forgiveness if, by the tears of his child, he is forced to commit perjury. No such concerns apparently troubled the historical Galileo; he had no wife, and his daughters were safely ensconced in convents. I think that modern audiences would not tolerate the sentimentality of this play. Ironically, in a play about censorship and misuse of power, the French government suppressed Scene 3 of Act II at the Paris premier. After a scene in which the Grand Duke of Tuscany advises Galileo that he is powerless to prevent extradition to Rome, Galileo ruefully contemplates the futility of relying on the power of princes, a pessimism not shared by the government of the Second Empire. The play ends with Galileo stamping his foot and saying, *"Et pourtant elle tourne!"* (And still it turns!), the same phrase which concludes the Stavis play. Again a historical inaccuracy, since Galileo never said, *"Eppur si muove."*

Brecht, by being daring and unconventional, has achieved a place for his play in many a repertory company. Theatrical companies rarely perform the other two plays. Brecht's Galileo is not an exact representation of the seventeenth-century scientist; instead he is a scientist created in our own image, and thus relevant to the needs of our own age.

In spite of Brecht's attribution of cowardice to Galileo, one gets a favorable impression of him after reading these plays, and most commentators have sided with Galileo. If Galileo is not a wholly admirable figure, the church is even less admirable. Not all authors have felt so well disposed toward Galileo, however. In particular, Arthur Koestler and Lewis Mumford have written negatively about him, and their views will be briefly outlined. In Chapter 5 of *The Sleepwalkers* Koestler removes Galileo from the pedestal where science mythology has placed him. Galileo brought all of his troubles upon himself, according to Koestler, because of his quasi-pathological contempt for his opponents, his suicidal folly in rejecting the Bellarmine compromise, and his patently dishonest answers to the questions of the Inquisitors. Koestler's bias against Galileo also surfaces in sympathy for Kepler in the Galileo-Kepler correspondence. Lewis Mumford's objections to Galileo are somewhat more abstract. He deplores the depersonalized world he sees around him at present. In Chapter 3 of the *Pentagon of Power* he blames Galileo for establishing the machine as the ultimate model of human thought. Galileo's "crime," according to Mumford, is trading the totality of human experience for the quantifiable variables, mass, length, and time. Galileo's error is thus the reverse of the Church Fathers who neglected study of the natural world in favor of the human soul. Koestler and Mumford represent minority opinions among Galileo critics, however.

J. Robert Oppenheimer

A Scientific Profile. I compared Galileo with his contemporary, Shakespeare; J. Robert Oppenheimer has a literary parallel with Gertrude Stein. Both Oppenheimer and Stein were more famous as catalysts, triggering creative reactions in others, than as creators themselves, although both produced some creative work. Oppenheimer, after studying under Lord Rutherford at the Cavendish Laboratories of the University of Cambridge and under Max Born at Göttingen, returned to America and divided his time between the University of California at Berkeley and the California Institute of Technology. Whereas the exciting development of the quantum theory in the twenties has been a European achievement, now the first American center for the study of theoretical physics coalesced around Oppenheimer. A brilliant physics professor who moreover could read Plato in the original Greek and quote the *Bhagavad-Gita* could not fail to develop a certain student following. Many of his students even imitated his manner of speaking, and asked themselves how Oppenheimer would attack a problem before they would attempt it. Stein, too, had held a "salon" in her Paris apartment. She recognized talent in Pablo Picasso, Ernest Hemingway, F. Scott Fitzgerald, Sherwood Anderson, and others, and encouraged them.

Oppenheimers's contributions to physics are not negligible. To him must go the credit for making the first prediction of the existence of the elementary particle known as the antiproton (a particle with the same mass as a proton, but opposite charge). He found the first example (field emission of electrons) of a process in physics known as "tunneling." A mechanical analogue would be an oil pipeline which tunnels through a hill. Similarly, charged particles can sometimes find their way through, rather than over, a potential barrier. Oppenheimer also theoretically explored gravitational collapse and neutron stars, but did not live to see these theoretical predictions verified by observational astronomy.

Although a thick curtain of secrecy still surrounds his wartime work on the Manhattan Project, enough is known to suggest that his role there was less original than managerial, as his suggestions and guidance helped many projects come to fruition. Most of the fundamental discoveries in nuclear fission had been made before the Los Alamos Laboratory was established; subsequent engineering work led to the successful atomic bomb test at Alamogordo. His quickness in comprehending the most tangled arguments, and his brilliance in exposition earned him a reputation as one of the country's leading scientists, but ultimately he lacked the degree of creativity which would have enabled him to launch sucessful and sustained attack on fundamental problems. Similarly, Stein has produced a number of successful literary efforts, among which are the play, *Four Saints in Three Acts* which Virgil Thomson turned into an opera, and the so-called *Autobiography of Alice B. Toklas.* She felt her experimental prose was better than that of James Joyce, a debate I will leave to the literary critics.

Fig. 11. J. Robert Oppenheimer (Photo: Los Alamos Scientific Laboratory)

When Oppenheimer stepped down in 1945 as Director of Los Alamos, he probably intended to devote more time to teaching at Berkeley and Pasadena, but this was not to be. Because of his reputation as the "father of the atomic bomb," his advice was eagerly sought in the centers of power in Washington. He became a member of dozens of government committees and Chairman of the General Advisory Committee, a group of nine leading scientists assembled to guide the Atomic Energy Commission. In October 1949 this committee met to debate whether the United States should undertake the development of a thermonuclear bomb. They did not doubt the eventual technical feasibiblity of building this awesomely destructive weapon, but they found a paucity of military targets (only the two largest Soviet cities) and felt that the United States' moral stance would be impaired if it were the first country to develop such a weapon. Their report[4] concluded, "In determining not to proceed to develop the superbomb we see a unique opportunity of providing by example some limitations on the totality of war and thus of eliminating the fear and arousing the hope of mankind." Oppenheimer was later to pay a high price for his lack of enthusiasm for this project.

After the committee's unfavorable report, a period of vicious infighting ensued between big-bomb advocates (Edward Teller, Lewis Strauss, and others) and opponents (Hans Bethe, Norbert Wiener, and others). The proponents readily enlisted the support of the Air Force and the Chairman of the Joint Chiefs of Staff, but the balance of power shifted decisively in their favor with the revelation of Klaus Fuchs' passing of atomic secrets to the Soviet Union. The General Advisory Committee reversed its decision of three months before and recommended to the President a crash program to develop the H-bomb.

When Edward Teller discovered in June, 1951, a "gimmick" which would make the H-bomb practical, virtually all opposition disappeared. Oppenheimer later recalled, "...when you see something that is technically sweet you go ahead and do it and you argue about it only after you have had your technical success."

Under the Eisenhower administration, several events took place which set the stage for Oppenheimer's downfall. The Soviet Union became the first nation to explode a thermonuclear weapon of the "dry" type (i.e., one using an isotope of the element lithium). At the same time Senator Joseph McCarthy was holding hearings in Washington to ferret out supposed Communists who had infiltrated the United States' government. Oppenheimer, in a series of brilliant speeches and articles, began to put into words the qualms that some of the nuclear pioneers at Los Alamos and elsewhere had felt. Then in 1953 another kind of "bomb" dropped. Lewis Strauss, the head of the Atomic Energy Commission, read to Oppenheimer a list of twenty-four charges against him, which, if proven, would deprive him of his security clearance, thus reducing his influence in government committees abruptly to zero. The first twenty-three charges covered mostly

[4]Writing in 1976 on the basis of the recently declassified General Advisory Committee's report on the H-bomb, as it came to be called, Herbert F. York concludes that the wisdom of the committee's advice has been confirmed by history. As Oppenheimer had recommended, American national security at that time did not require the immediate initiation of a high-priority program to develop a hydrogen bomb. It was a unique opportunity to use the first Soviet nuclear explosion as a lever to begin arms-control talks before weapons of a more destructive character were devised. Our foregoing the development of a hydrogen bomb was a necessary, but not a sufficient, condition for other nations to abstain from weapons development.

former Communist associations, old accusations which had been thoroughly sifted many times before by security agents who had never seen fit to deny him a security clearance. The last charge was a new one and proved his undoing. It read in part as follows: "...even after it was determined, as a matter of national policy, to proceed with the development of a hydrogen bomb, you continued to oppose the project and declined to cooperate fully in the project."

Basing its decision primarily on the damning testimony of Edward Teller, the Personnel Security Board ruled against Oppenheimer. Of the three-member board, only Ward V. Evans voted in favor of Oppenheimer. Evans noted that most of the charges against Oppenheimer concerned a time before his service at Los Alamos. He felt that Oppenheimer had honestly given his best professional advice when asked to do so by the General Advisory Committee. Was it wrong, he asked, to have moral and ethical reservations about a weapon whose deployment would have unknown ramifications?

Oppenheimer's reputation among fellow scientists has fluctuated, reaching its high point in the period immediately following his security hearing in 1954. His shabby treatment by the review board spotlighted his dissenting opinions on military strategy, making him a hero to liberal scientists. Nevertheless, as chief scientific advisor to the Truman administration, he had not been an open critic of American weapons policy. Other scientists, such as James Franck, I.I. Rabi, Edward U. Condon, and Leo Szilard, criticized government policy in the pages of *The Bulletin of the Atomic Scientists,* but Oppenheimer maintained public silence until the Republicans took office in 1953. Nor is Oppenheimer's private record unblemished. He volunteered hearsay evidence to security agents concerning Communist associations of friends and students, including Bernard Peters, Rossi Lomanitz, and David Bohm. Because of Oppenheimer's testimony, these men lost their jobs at leading universities, and received far less publicity in the scientific community that Oppenheimer did at his hearing. He gave exaggerated importance to Haakon Chevalier's mentioning that Chevalier knew a man who sought scientific information for the Soviet government. After Oppenheimer, under pressure, reported the incident to security agents, Chevalier lost his job and could find no further employment in the United States. Oppenheimer later admitted that he had been "idiotic" to act in such a way. Thus Oppenheimer had cooperated with the security system in the past and accepted its legitimacy. Can this explain why he was curiously passive and acquiesced in his own destruction at the hearing?

The political climate again shifted, when in 1958 the United

States began to discuss a test-ban treaty with the Soviet Union. When President John F. Kennedy was assassinated, the Enrico Fermi award, the nation's highest honor in nuclear science, was awaiting his signature. President Lyndon Johnson signed the award and presented it to J. Robert Oppenheimer for services rendered to the atomic energy program in its crucial years.

On the Stage. Heinar Kipphardt has taken the Atomic Energy Commission's report, *In the Matter of J. Robert Oppenheimer*, as the basis for his documentary drama of the same title concerning the Oppenheimer security hearing. By selecting material from the three thousand or so pages of this document, he has not only given dramatic cogency to the hearing, but also added a punch that only stark truth can have. Nevertheless, his method has its limitations. Roger Robb, counsel to the A.E.C., stands out in the play as the obvious villain. He functions as a prosecuting attorney, a master of innuendo, half-truth, and implication. Yet he is only acting for the chairman of the A.E.C., Lewis L. Strauss. Since Strauss didn't attend the hearings, he does not appear in the proceedings.

Oppenheimer wrote to Kipphardt protesting the production of the play. He maintained that the actual proceedings were a farce, and that Kipphardt was trying to make them into a tragedy. He felt that the improvisations introduced by Kipphardt were distorting the factual record and suggested that it would be more appropriate for a German author to seek material for his plays in the sequence of events implied by the cities Guernica, Coventry, Hamburg, Dresden, Dachau, Warsaw, and Tokyo. In addition, he threatened legal action. As a result of this letter from Oppenheimer, Jean Vilar deleted the screen projections and monologues from the Paris premier of the work.

While Oppenheimer does have a point in that the play is not strictly historical, the distortions seem to have been minor. Oppenheimer spent the night with his former communist fiancée, Jean Tatlock, not in a hotel as the play has it, but in her parents' home. Kipphardt gives Oppenheimer two defense attorneys, whereas Oppenheimer actually had three. Similarly, in the play there are six witnesses; in the actual hearing there were about forty. In the play the board announces its findings at the end of the hearing and gives Oppenheimer the opportunity of a last speech; in reality the board did not announce its findings at the hearing, and Oppenheimer used his opportunity only to make a technical point.

In the parry and thrust of forensic debate many questions emerge. What kind of people are physicists? Is there such a thing as ideological treason? How much freedom should be sacrificed to the security system which is set up to protect freedom in the first place? Is there such a thing as guilt through association? Why should supposedly scientific decisions take on different colorations in different times? If anything, the author has given the playgoer too many themes with which to cope in one evening of the theater. In this respect the play merely reflects the complexities of a real-life situation.

The central theme seems to be Kipphardt's vision of Oppenheimer as a twentieth-century Galileo, sacrificing his moral integrity to the military in order to pursue scientific interests. In doing so, the physicists "have known sin, " to use the Biblical phrase Oppenheimer applied to the situation. To place the military defense of one nation ahead of the safety of mankind as a whole was short-sighted. The climax of the play comes in the testimony of Edward Teller, who says that he would feel more secure with the vital interests of the country in other hands. Teller expressed this view in other speeches and writings as, for example, in his book, *Legacy of Hiroshima.* He argues there that the country desperately needed the hydrogen bomb for its defense, and could have had it four years sooner if Oppenheimer had put his influence whole-heartedly behind its development.

Galileo and Oppenheimer Compared

The sensational Oppenheimer affair corresponds in many ways to the dramatic events surrounding Galileo. One could hardly pick two more dramatic figures from the pages of the history of science than Galileo Galilei and J. Robert Oppenheimer. G. de Santillana has summarized the many similarities in their careers. Both were useful to society and "delivered the goods"; the telescope and the geometric and military compass in the case of Galileo, and the atomic bomb in the case of Oppenheimer. Galileo obtained an *imprimatur* for his book, *Dialogues Concerning the Two Chief World Systems,* and Oppenheimer had a security clearance, but both ran into trouble when they tried to exert influence in matters of high policy. The "crime" was lack of proper enthusiasm for church directives in the case of Galileo, for security directives in the case of Oppenheimer. In each instance the scientist threatened a monopoly: Galileo's scientific publication in Italian made him a competitor of the Jesuits in education, while Oppenheimer's emphasis on tactical nuclear weapons threatened the Air Force's vested interest in strategic weapons deliverable by plane. During their trials (more accurately a "security hearing" for Oppenheimer) both scientists were shown a great deal of official consideration. The prosecution skirted the main issues; Copernican theories did not come up for discussion at Galileo's trial, nor were technical matters concerning the hydrogen bomb brought up at Oppenheimer's, for the ostensible reason that his lawyers lacked security clearance. In each case the prosecution had to dig into the past for its charges; the Holy Office found an Injunction allegedly given to Galileo 17 years earlier, and Atomic Energy Commission board dug up the Chevalier incident from 11 years previous. Neither was convicted of major infamy (heresy for Galileo, treason for Oppenheimer), but in the eyes of the public the result of the trial amounted to the same thing: these scientists were either too clever or too scared to enter the big leagues. Both trials ended in public humiliation and recantation for the defendants. Oppenheimer in effect recanted at the beginning of his trial in the abject letter[5] of March 4, 1954. Both came to be viewed, rightly or wrongly, as martyrs to the cause of scientific truth, whether persecuted by Big Church or Big Government.

Albert Einstein

A Scientific Profile. In 1905, that *annus mirabilis* in the history of science, an obscure Swiss patent-office clerk named Albert Einstein published four papers in the prestigious German physics periodical *Annalen der Physik.* How did such a remarkable situation come about?

As a young man, Einstein had always hated military parades, whose participants he saw as ranks of robots. In school the memorization and repetition which characterized the German method of instruction of the time in the study of foreign languages, history, and other subjects struck him in the same way, and he refused to be herded with the other students through these drills. But he studied mathematics and physics on his own, attaining a level of competence far beyond the other students. As a consequence of his one-sided education he failed the entrance examination for the Federal Institute of Technology in Zurich, necessitating further study before he could gain entrance. After graduation he sought unsuccessfully to become a research assistant to some of the more famous professors in Germany and the Netherlands. Finally, through the intervention of a friend, he secured a job in the patent office at Bern, enabling him to work on important problems in physics in his spare time.

A comparison with Newton is suggestive: both made important contributions to science in a quiet, secluded atmosphere

[5]Joseph Boskin and Fred Krinsky have produced a book entitled *The Oppenheimer Affair: a Political Play in Three Acts* (1968). Although the book takes the form of three "acts" plus "critics' reactions," it is not really a play, but it does make a useful companion volume to *In the Matter of J. Robert Oppenheimer.* The Boskin and Krinsky book reprints the charges against Oppenheimer, as well as Oppenheimer's reply of March 4, 1954, amounting to a confession. The Lansdale and Teller testimonies and the findings of the Personnel Security Board are presented at greater length than in the play. Finally, seven editorials and magazine articles giving reactions to the Oppenheimer affair are reprinted.

when they were young men: Newton at Woolsthorpe when he was twenty-three, and Einstein in the patent office when he was twenty-six. Furthermore, Isaac Barrow resigned his professorship at Cambridge in favor of Newton, and Friedrich Adler refused to accept the newly created position of *Professor Extraordinarius* at Zurich University, saying that the university could ill afford to pass up someone with the scientific genius of Einstein. They both pondered the nature of light and the laws of motion. In later years both Newton and Einstein became public figures.

Volume *17* of *Annalen der Physik* contained the four papers which almost overnight made Einstein world famous. The first set out his explanation of the Photoelectric Effect which he envisaged as taking place in collisions between light *particles* (photons) and electrons. Entitled, "On a Heuristic Point of View Regarding the Production and Transformation of Light," it explained light by invoking particle-like properties, and deliberately ignored other experiments (diffraction of light by a slit, for example) that could be explained only by wave properties of light. The next paper dealt with Brownian Motion, the random motion of macroscopic particles, such as pollen, suspended in a liquid and continually hit by the liquid's moving, invisible molecules. The Frenchman Jean Perrin later verified Einstein's theory in a series of brilliant experiments providing direct evidence for the existence of atoms.

Einstein's third paper, "On the Electrodynamics of Moving Bodies," launched his Theory of Special Relativity. His acute questioning of the Newtonian assumption of an absolute space and time reference frame led him to some strange conclusions. He discovered a world in which mass increased with the speed of a body, lengths contracted, and time intervals became longer. These effects became more pronounced as the speed of a body approached the speed of light, a speed which could not be exceeded. Since then, the Theory of Special Relativity has been abundantly verified, but at the time its revolutionary character prevented immediate acceptance. In awarding him the Nobel Prize in 1922 the Swedish Academy conservatively avoided any mention of relativity; instead, they gave it to him for the discovery of the Photoelectric Law. But this did not stop Einstein from giving his acceptance speech on the subject of relativity! To have written any one of these major papers of 1905 would have made him famous, but to have made three such contributions gives him the stature of a genius such as Newton.

The fourth paper was also on the topic of special relativity. He answered affirmatively the question, "Does the inertia of a body depend upon its energy content?" He showed that if a body gave off energy E in the form of radiation, its mass diminished by E divided by the speed of light squared. Later generalized in the famous relationship, $E=mc^2$, it implied that all matter had a huge inherent energy content which could be liberated under the proper circumstances. The mushroom cloud over Hiroshima was abundant evidence of the truth of Einstein's simple equation. Ironically, the mild-mannered violinist, a pacifist until Belgium was in danger of being overrun at the beginning of World War II, laid the groundwork for this awesomely destructive weapon.

During the period 1913-1933 Einstein was a professor at the University of Berlin and an associate of the Kaiser Wilhelm Institute. A physics colloquium at the University of Berlin might draw such luminaries as Max Planck, Walter Nernst, Max von Laue, and Lise Meitner. Out of this heady atmosphere came Einstein's 1916 paper on the Theory of General Relativity. In it he applied the principle of relativity to gravitation, using a branch of mathematics known as "tensor analysis," and thereby made it even more inaccessible to the non-scientific intellectual than the Special Theory of Relativity. Einstein suggested ways in which the theory could be checked experimentally, and the difficult experiments were performed with positive results. Astronomers had noticed that the perihelion (point of closest approach) of the orbit of the planet Mercury was advancing by about 5600 seconds of arc per century. Although mostly explained by Newtonian physics, a discrepancy of about 45 seconds of arc remained in the forward direction. Einstein's theory predicted an advance of about 43 seconds of arc, thus neatly completing the understanding of planetary motions.

Moreover, the General Theory of Relativity predicted two effects previously totally unknown, the bending of starlight as it passed close to a massive body, and the change in wavelength of light emitted by a star. Both effects were found in subsequent years, the first in a quite dramatic manner. The Royal Society and the Royal Asstronomical Society of London appointed a committee headed by Sir Arthur Eddington to prepare an expedition to observe the solar eclipse of March 29, 1919. In case inclement weather during the 302 seconds of totality should forestall the taking of photographs, two parties were sent out, one to Sobral, Brazil, and the other headed by Eddington himself, to the island of Principe in the Gulf of Guinea. Highlighting the international character of science, British scientists were trying to validate the theory of a scientist from another country, recently an enemy. The photographs unmistakably showed starlight bending closer to the sun when the light passed by at grazing incidence.

Later in the year the two societies held a joint meeting to announce the dramatic confirmation. The mathematician and philosopher, Albert North Whitehead, gave the following eye-witness account:

> It was my good fortune to be present at the meeting of the Royal Society in London when the Astronomer Royal of England announced that the photographic plates of the famous eclipse, as measured by his colleagues in Greenwich Observatory, had verified the prediction of Einstein that rays of light are bent as they pass in the neighborhood of the sun. The whole atmosphere of tense interest was exactly that of the Greek drama. We were the chorus commenting on the decree of destiny as disclosed in the development of a supreme incident. There was a dramatic quality in the very staging—the traditional ceremonial, and in the background the picture of Newton to remind us that the greatest of scientific generalizations was now, after more than two centuries, to receive its first modification. Nor was the personal interest wanting; a great adventure in thought had at length come safe to shore.
>
> . . . . .
>
> The essence of dramatic tragedy is not unhappiness. It resides in the remorseless working of things . . . This remorseless inevitableness is what pervades scientific thought. The laws of physics are the decrees of fate.

Another paper of the Berlin years introduced the concept of *stimulated emission* of photons, a process later utiltzed in the device now known as the laser (an acronym for Light Amplification by Stimulated Emission of Radiation). The Berlin group was later joined by the Austrian physicist, Erwin Schrödinger, who in 1926 devised a theory, later known as *wave mechanics,* that described the wave properties of matter. According to an interpretation of this theory, principally by the so-called Copenhagen School (Niels Bohr and collaborators), future observable events can be predicted only in a statistical way. Despite a brilliant series of successful applications to the atomic realm, the proponents of the new theory failed to convince Einstein, and he engaged Bohr in a continuing debate concerning the epistemological and philosophical bases of wave mechanics. That the universe should operate in this manner went against Einstein's intuitive feelings, occasioning his famous remark that "God does not play dice." Ironically, this kind of bias had prevented early acceptance of his relativity theories by many scientists.

As the Second World War came on, Einstein found a haven at the Institute for Advanced Study at Princeton. He unsuccessfully tried to find a unified field concept that would explain gravitational, electromagnetic, and atomic phenomena. Meanwhile, an ominous event occurred: Germany forbade the export of uranium from mines which she controlled in Czechoslovakia. This act, combined with the knowledge of the previous discovery of uranium fission in Berlin in 1938 by Hahn and Strassmann, made

Fig. 12. Einstein riding his bicycle at the California Institute of Technology.

it seem likely that the Germans might try to make a uranium bomb. At the urging of Leo Szilard and Edward Teller, Einstein wrote a letter on August 2, 1939 to President Roosevelt calling to his attention the potential of a uranium chain-reaction bomb. A second, more urgent letter followed in early 1940. After the war, Einstein worked for arms control, and two days before his death in 1955, in what may have been his last public act, he put his signature on a statement drafted with his approval by Bertrand Russell about the likely nature of future wars. This gave impetus to a series of meetings called "Pugwash Conferences" devoted to discussion of nuclear arms control.

At his own request no monument was erected over his remains; he had no need of one, since his theory of relativity stands out as one of the glories of human intellectual achievement.

On the Stage. Einstein's life has inspired the authors of a variety of stage works; Friedrich Dürrenmatt has written a play with Einstein (or facsimile) as a central character, composer Paul Dessau and librettist Karl Mickel have written a biographical opera, and Philip Glass and Robert Wilson have staged a multimedia "event" which at least invoked the aura of Einstein. Each of these will be considered in turn.

Friedrich Dürrenmatt's play, *The Physicists* (1962), postulates a reverse reality in which the world is a vast madhouse, and only in a sanitarium can sane men find refuge. Dürrenmatt counterbalances the theme of madness by observing the Aristotelian unities: the play is confined to a single action, occurring in a single place, occupying only a day's time. As the play opens, the patient who thinks he is Einstein has strangled his nurse. This is the second murder; three months earlier another nurse was strangled by another patient masquerading as Newton. One questions the innocence of the sanitarium director, Fraülein Doctor Mathilde Von Sahnd, when she reveals that she is keeping her great-aunt under continuous sedation. Nurse Monika falls in love with a third physicist, Johann Wilhelm Möbius, whom she believes to be sane, although he has visions of King Solomon. She urges Möbius to leave the sanitarium under her care; he responds by warning her to flee the sanitarium alone. Monica refuses to listen to him, and Act I ends as Möbius strangles her.

In Act II "Newton" and "Einstein" are revealed as "real" physicists working as secret-service agents from the West and East, respectively; they have been sent to track down the famous physicist Möbius and bring him back. But Möbius has foiled them by burning his papers so that his revolutionary "Unitary Theory of Elementary Particles" would not fall into the hands of either power block. They have all murdered because their nurses either stumbled onto the truth or wanted to take them away. In the ensuing deadlock, Möbius argues that the three of them must remain in the sanitarium to prevent the destruction of humanity. In this way the three deaths become sacrifices instead of ordinary murders. One wonders if real secret-service agents would agree to this solution. The third and final twist in the story is the doctor's revelation that King Solomon has long ago ordered her to make photocopies of all of Möbius' work. Her cartel will henceforth ransack the world; the three physicists are prisoners. The messages in this sardonic parable are clear: even a madhouse offers no refuge, and the competition between the two superpowers will eventually be exploited by unreasoning forces that neither anticipated.

Is the life of Albert Einstein a suitable subject for an opera? The answer is *yes* according to the composer Paul Dessau whose opera *Einstein* was premiered by the Berlin State Opera in 1974. Karl Mickel's libretto treats the social responsibility of the scientist in a Brechtian manner. Carrying Brecht's *Galileo* one step further, Mickel allows political considerations to dominate historical events and artistic choice. It was written in 1970-72 during the Vietnam War; if written today it might have a different political coloration. The main action is so loosely based on the life of Albert Einstein that it is more of a morality play than a biography. It incorporates at least the following historical events: Einstein's flight from Nazi Germany, his protest against Nazi aggression, and his letter to Roosevelt which alerted the president to the possibility of constructing an atomic bomb. The theme is that imperialistic elements in the White House and Pentagon make use of scientists for their own ends; individual good will and humanism are not adequate to resist these forces. According to Mickel, the politically naive are manipulated by the militarists, but the organized forces of the people (Communism) have a chance of overcoming the imperialists.

The action flows in two streams, each with its own distinct music, orchestra, and characters. Three intermezzi constitute one of the streams. They are farces involving Hanswurst, a traditional German folk-clown comparable to the English character Punch. A Prussian cop throws Hanswurst three times to a man-eating crocodile which symbolizes fascism and imperialism. Each time Hanswurst tells a joke so that the crocodile will open his jaws and laugh. The first time he escapes, the second time he is half eaten just at the moment in the main action when the atomic bomb goes off, and the third time he escapes over a giant razor blade. Hanswurst is a sacrificial lamb, but he always revives, dancing on the razor's edge to music which sounds like a cross between Viennese operetta and Wagner. He seems to represent the masses who may be oppressed but not defeated.

The main action concerns Einstein and two colleagues, a younger and an older one. Einstein realizes that he must flee Germany when his effigy is burned along with some of his books. Bach's music is played as SS men destroy Einstein's house, symbolizing the destruction not only of the house, but of the whole German humanistic cultural heritage. Einstein's colleagues fall into the hands of the Nazis; the younger one is thrown into prison, and the older one agrees to work on an atomic bomb for the Führer so that the younger one may be spared. In the United States, Einstein encounters the younger colleague after he has escaped from Germany and become an American officer. He is told about the older physicist who builds atomic bombs for Hitler using Einstein's writings. A series of duets with the famous scientists Giordano Bruno, Galileo, and Leonardo da Vinci follow as Einstein agonizes over his alternatives. The scientists from the past represent resistance to authority, abject submission, and silence about real discoveries, respectively. Einstein feels he cannot follow the advice thrown at him by Leonardo, "Da siehe du zu, " ("Just look on").

In the end Einstein goes to Roosevelt to offer his services for work on the bomb. In a tremendous theatrical climax, the atomic bomb explodes, the sun turns black, and Einstein becomes a hundred-year-old man. He realizes it was murder, and says, "I am death." Actually it was Oppenheimer, not Einstein, who said, "I am become death, the shatterer of worlds," a quotation from the *Bhagavad-Gita*. Einstein calls for resistance, is denounced by his two colleagues (the older one had been dragged out of Germany to help with the American atomic bomb), and is haled to the Supreme Court where he is "sentenced" to everlasting fame. Later, the younger colleague sees the light and presumably joins the Communist party. As the opera ends, Einstein, in seeking to avert a new disaster, burns one of his formulas. Yet he gives his book to a child, hoping that the human race may be wiser by the time the child is ready to use it.

Einstein's gift of his book to the child is accompanied by the music of Bach. In fact, throughout the main action musical quotations from other composers and from Dessau's previous work appear in a straightforward form or in parody and transformation. Mostly the work is in the 12-tone style, a technique prohibited for years in East Germany.

Another work, *Einstein on the Beach* (1976), by Philip Glass and Robert Wilson, is not really an opera, but a multi-media showpiece in which there are no singing characters. The name comes from a combination of the title of Nevil Shute's novel, *On the Beach,* and of Einstein himself. Einstein's violin sings in ironic counterpoint to images of nuclear holocaust. The music—repetitive, ritualistic, and incantatory—contrasts with the mechanized world of trains and spaceships depicted on stage. The message seems clear: magic and myth have more to offer than science and technology

The intellectual migration from Europe to America in the 1930's placed German exiles in a position to make a unique contribution to the struggle against German totalitarianism. America's first use of the atomic bomb against Japan is beside the point; the motivation for developing it was that fission had first been discovered in Germany, and Hitler's scientists were supposedly working on such a weapon. The questions which Dürrenmatt and Mickel raise are enduring ones. Did these scientists betray their ideals in any important way when they agreed to work on the atomic bomb? Are there areas of research in which scientists should refuse to work? Can scientific questions be considered apart from their political implications?

Johannes Kepler

On the Stage. Another scientist whose life has provided the material for an opera is Johannes Kepler, whose ideas have been outlined in Chapter 2 above. Paul Hindemith's *Die Harmonie der Welt* (1957) with libretto by the composer is an accurate portrayal of the last third of Kepler's life. The plot is not a romantic but an intellectual "eternal triangle"; Hindemith probes relationships between the scientist, society, and his family. Hindemith's broad canvas includes characters from Kepler's personal life (his step-daughter Susanna and second wife of the same name), religious life (the priest Hizler), political life (Emperors Rudolph II and Ferdinand II, Wallenstein), as well as his intellectual life (Kepler's assistant). To insure historicity, the composer-librettist has had to include a number of minor characters whose appearance detracts from a natural dramatic evolution of the action. This category includes Kepler's brother Christoph and Baron Starhemberg, who raised Kepler's second wife.

Kepler's personal triumph and tragedy is at times submerged in the turbulent events of the Thirty Years' War. In this respect the opera is like Moussorgsky's *Boris Godunov,* in which Boris is quite often not the center of attention. Typical of this dramatic weakness is the death scene of Kepler in Regensburg, accurate to a fault. The lights come up on a split-level stage. In one room electors are forcing Emperor Ferdinand II to dismiss General Wallenstein. They complain that the war has already lasted 13 years and that Wallenstein's megalomania threatens to consume them all. The feverish Kepler lying in bed in another room hears and sees these actions as in a dream. Kepler did actually die of a fever, and at the last he couldn't speak but merely pointed to the sky. Kepler's poignant gesture is bound to be overshadowed by the emperor in full regalia surrounded by his Catholic princes.

The opera contains little old-fashioned face-to-face human conflict, long a staple of the operatic stage. An exception to the general trend of the opera is a scene in the Second Act. The zealot Daniel Hizler confronts Kepler at the church door at Linz, and denies him communion until he renounces his heretical views. Kepler did not accept the Lutheran doctrine of the omnipresence in the world of not only the spirit but the body of Christ, so he could not sign the Concordat formula. The acceptance of this formula was not universal; only Mecklenburg and Württemburg ("Lutheran Spain") adopted it.

Another story the libretto follows is that of Kepler's mother Katerina. In the First Act she is in a cemetery in Württemberg looking for her father's grave. She wants to find his skull and have it set in gold to make a drinking cup! According to a superstition, drinking from such a vessel will protect one from spiritual arrogance (she has Kepler in mind). Four women see her and also remember that she touched four cows and their milk turned sour. She is accused of witchcraft and flees to Kepler's house in Linz. In a stroke of dramatic genius Hindemith has Katerina and Kepler's wife Susanna on stage at the same time in separate rooms unbeknownst to each other quoting antagonistic Biblical texts. Obviously, this quarrelsome old woman will not be welcome for very long in the household. She leaves and is eventually brought to trial for witchcraft. As the instruments of torture are brought out, Kepler enters and succeeds in getting her acquitted.

As if this weren't enough for one opera, the libretto closely traces the background of political events. It follows the rise of General Wallenstein for whom Kepler casts a horoscope. Wallenstein's obsessive drive for power is abetted by the opportunistic Tansur (a non-historical character whose name is an anagram for Saturn). Tansur is first seen on the streets of Prague peddling prints of the great comet and saying it augurs disasters ahead. When Kepler's assistant gives a scientific explanation for the comet, Tansur becomes angry because it may drive away customers. He later becomes Wallenstein's majordomo. At the height of his power Wallenstein employs Kepler, who retains his title as Imperial Mathematician. They are working at cross purposes; Wallenstein wants Kepler's aid in calculating logistics and weather, duties incompatible with the abstract nature of Kepler's thought. Kepler explains that pure science is not immediately useful.

The Fifth Act concludes in a mystical apotheosis. After Kepler's death the actors reappear in costumes of baroque splendor suggesting the astrological significance of the sun, its six nearest planets, and the moon. Kepler plays the role of the earth, Katerina the moon (appropriately enough considering Fiolxhilde's association with lunar spirits in *Somnium* as we have seen in Chapter 2), and Emperor Ferdinand II the sun. The other planets are represented by Hizler as Mercury, Susanna as Venus, Kepler's assistant as Mars, Wallenstein (appearing after his murder) as Jupiter, and Tansur as Saturn. Zodiac signs and a chorus representing the Milky Way appear. The harmony of the world is only attainable in the celestial sphere. The chorus asserts that dreams, intuition, and prayer are higher than knowledge, exploring, and learning, a conclusion that Kepler might well have agreed with.

Inevitably, the number-mysticism of Kepler suggested itself to Hindemith as a subject for a libretto. Hindemith sees his harmonic relationships as founded on natural laws rather than arbitrary rules, a weakness of the 12-tone system. Kepler, too, sought simple numerical relationships, but only those read from the book of nature were valid for him.

The evidence suggests that having a plot with "meat" in it, that is, a highly intellectual libretto such as this, is irrelevant to the needs of opera. Audiences tolerate many absurdities in an opera with a strong vocal line; without it an opera suffers no matter how carefully crafted the plot. Critics have characterized the choral writing of this opera as gray and colorless and complained that the solo writing inadequately differentiates the various characters. Considered vocally, Hindemith's *Harmonie* is not an unqualified success.

A three-movement symphony drawn from the music of the opera has gained acceptance by many audiences. Similarly, Hindemith has produced a symphony from his earlier opera, *Mathis der Maler,* a work also treating the theme of a historical character influenced by violent events. Given the relative popularity of these symphonies and the dramatic weaknesses of the operas, it would seem that this composer is more at home in the symphonic than the operatic medium.

The twentieth-century stage works considered in this chapter often depict the scientist battling, and to some extent controlled by, large institutions. Galileo battles the dogma of the church; Oppenheimer, the Atomic Energy Commission and indirectly the Air Force. According to the librettist of *Einstein,* the Pentagon uses the scientist for its imperialistic purposes. In contrast, the works of early science fiction in the previous chapter emphasize individual scientists controlling nature or manipulating human life. The stage works do not provide unequivocal answers to the problems of institutional control of science. Galileo still makes a significant contribution to science in spite of official condemnation, but Oppenheimer does not. Kepler's exclusion from communion and his mother's trial for witchcraft almost succeed in deflecting him from his scientific pursuits. In certain segments of society, the solution of Hindemith at the conclusion of *Die Harmonie der Welt* is currently popular: a personal and anti-intellectual retreat into mysticism.

Bibliography

Boskin, Joseph and Krinsky, Fred, *The Oppenheimer Affair: A Political Play in Three Acts* (Glencoe Press, Beverly Hills, 1968).

de Broglie, L., Armand, L., Simon, P.-H., *Einstein* (Peebles Press, New York, 1979).

Chevalier, H., *Oppenheimer, The Story of a Friendship* (G. Braziller, New York, 1965)

____. *The Man Who Would be God.*

Clark, Ronald W., *Einstein: The Life and Times* (World Pub. Corp., New York, 1971).

Davis, N.P., *Lawrence and Oppenheimer* (Simon and Schuster, New York, 1968).

Dukas, H. and Hoffman, B., *Albert Einstein: The Human Side* (Princeton Univ. Press, 1979).

Frank, Philipp, *Einstein: His Life and Times* (Alfred A. Knopf, New York, 1972).

French, A.P. (Ed.), *Einstein: A Centenary Volume* (Harvard Univ. Press, Cambridge, 1979).

Geymonat, L., *Galileo Galilei* (McGraw-Hill, New York, 1957).

Hoffmann, Banesh, *Albert Einstein: Creator and Rebel* (Viking Press, New York, 1972).

Koestler, Arthur, *The Sleepwalkers* (Macmillan, New York, 1959).

Major, John, *The Oppenheimer Hearing* (Stein and Day, New York, 1971).

Mumford, Lewis, *The Myth of the Machine*, Vol. II, *The Pentagon of Power* (Harcourt Brace Jovanovich, New York, 1970).

Panofsky, Erwin, *Galileo as a Critic of the Arts* (M. Nijhoff, The Hague, 1954).

Rabi, I.I., et al., *Oppenheimer* (Charles Scribner's Sons, New York, 1969).

Ronan, Colin A., *Galileo* (G.P. Putnam, New York, 1974).

Rouzé, Michel, *Robert Oppenheimer: The Man and His Theories* (P.S. Eriksson, New York, 1965).

de Santillana, G., *The Crime of Galileo* (Univ. of Chicago Press, 1955).

____. "Galileo and J. Robert Opoppenheimer", The Reporter, **17**, 10, 1957

Schroeer, Dietrich, "Brecht's *Galileo:* A Revisionist View, " Am. J. Phys., **48**, 125, 1980.

Seeger, R.J., *Men of Physics: Galileo Galilei* (Pergamon, Oxford, 1966).

Shepley, J.R., and Blair, C., Jr. *The Hydrogen Bomb* (David McKay, New York, 1954).

Stern, P.M., *The Oppenheimer Case* (Harper and Row, New York, 1969).

Teller, Edward, with Brown, Allen, *The Legacy of Hiroshima* (Doubleday, Garden City, 1962).

Ulam, S.M., *Adventures of a Mathematician* (Charles Scribner's Sons, New York, 1976).

USAEC, *In the Matter of J. Robert Oppenheimer*, Transcript of hearings before the Personnel Security Board, (U.S. Gov't Printing Office, Washington, D.C., 1954).
York, Herbert F., *The Advisors: Oppenheimer, Teller and the Superbomb* (W.H. Freeman, San Francisco, 1976).

Plays and Libretti

Brecht, Bertolt, *Galileo* (Grove Press, New York, 1966).
Čapek, Karel, *R.U.R. (Rossum's Universal Robots)* (Oxford Univ. Press, 1923).
Dürrenmatt, Friedrich, *The Physicists* (Grove Press, New York, 1964).
Hindemith, Paul, *Die Harmonie der Welt* (B. Schott's Söhne, Mainz, 1957).
Kipphardt, Heinar, *In the Matter of J. Robert Oppenheimer* (Hill and Wang, New York, 1968).
Ponsard, François, *Galilée in Oeuvres Complêtes* (Michel Lévy Frères, Paris, 1876).
Rice, Elmer, *The Adding Machine* in *Three Plays* (Hill and Wang, New York, 1965).
Stavis, Barrie, *Lamp at Midnight* (A.S. Barnes, New York, 1966).

Chapter 4

Science In English Poetry

THE "NEW PHILOSOPHY" of Copernicus, Kepler, and Galileo made slow progress in the seventeenth century. Hardly anyone before Newton had read Kepler except Gassendi, and, surprisingly, the poet John Donne. Galileo's trial had put the clamps on free discussion of Copernican ideas in Roman Catholic countries. In spite of these obstacles, the Ptolemaic, Copernican, and Tychonic theories became known in intellectual circles. In addition an older idea, the daily rotation of the earth, often figured in literary discussions. At the beginning of the seventeenth century it was unclear which theory would triumph, or what difference it would make in everyday affairs. A few poets caught intimations of a new temper of thought. Percy Shelley says in his *Defense of Poetry* that poets are "the mirrors of the gigantic shadows which futurity casts upon the present." In another scientific metaphor, poets are barometers registering changes in the intellectual climate before these changes are perceived by the majority of people. John Donne was one poet who had such a gift; his mixture of faith and skepticism was a breath of fresh air ushering in the modern era.

John Donne

John Donne was born into a Roman Catholic family, received early training from the Jesuits, and university education at Oxford and Cambridge. In later life he became a Protestant, rising to the post of Dean of St. Paul's Cathedral. In 1611 Donne probably had no idea that he would become a clergyman of the Church of England, although his evolution toward Protestantism had progressed rather far. In that year (only one year after Galileo's *Sidereus Nuncius* and not long after Kepler's *De Motibus Stellae Martis)* Donne published the satire, *Ignatius his Conclave,* in order to air his grievances against the Society of Jesus. Notable today more for the revelation it gives of Donne's reactions to the "New Philosophy" than for its attacks on the Jesuits, *Ignatius his Con-*

clave required of the author an act of poetic imagination to envision the conditions under which Satan would grant a coveted place in his inner room, much as Kepler's *Somnium* required an act of geometric imagination to visualize the motions of the earth and sun as seen from the moon. Manuscripts of *Somnium* circulated in England as early as 1610, and it is likely that Donne knew of them. Indeed the first words of *Ignatius his Conclave* suggest a cosmic voyage beginning with a trance just as in *Somnium*.

> I was in ecstasy, my little wandering sportful soul, guest, and companion of my body had liberty to wander through all places.

But Donne drops this device in the main body of the work, only to resurrect it in the concluding sentences, "And I returned to my body."

Donne's association with Kepler is stronger than mere literary parallels. Donne and Kepler actually met, though several years after Donne wrote *Ignatius his Conclave*. Donne accompanied a diplomatic mission undertaken by the Earl of Doncaster in 1619 at the behest of King James I. In the course of their travels through Germany and Austria to initiate discussions with the Holy Roman Emperor, Doncaster and Donne arrived at Linz where Kepler lived. Kepler wrote in an undated letter to an unnamed correspondent that he had met a "Doctore Theologo Namens Donne" who had arrived on October 23. The subject of the letter is the presentation to James I of one of Kepler's books, which Wilbur Applebaum believes is Kepler's *Harmonice Mundi (Harmony of the World)*. The letter contains no clues as to what other subjects Donne and Kepler might have discussed.

Another link between Donne and Kepler is a book by Kepler, *Eclogae Chronicae* (1615), discovered in Donne's library. The book consists of letters to Sethus Calvisius, Markus Gerstenburger, Johannes Deckers, S.J., and Herwart von Hohenburg, in which Kepler details his side of the controversy over certain dates in the life of Christ. It reflects Donne's interest in science and religion, but it lies in a side branch rather than the mainstream of contemporary astronomical debate. Thus Donne was probably aware of some of the developments in astronomy without fully understanding the technicalities of competing theories or the importance of the scientific methods from which they arose.

In *Ignatius his Conclave* the great innovators—Copernicus, Machiavelli, Paracelsus, Columbus, and others—are haled before Satan and Loyola in Hell and charged with having given

> an affront to all antiquity, and induced doubts, and anxieties, and scruples, and after, a liberty of believing what they would; at length established opinions directly contrary to all established before.

Loyola's reply after Copernicus has stated his case shows that when Donne reflected scientifically upon the Copernican system he considered it superior to the more ancient Ptolemaic system. Loyola addresses Copernicus as follows:

> What cares he [Lucifer] whether the earth travel, or stand still? Hath your raising up of the earth into heaven, brought men to that confidence, that they build new towers to threaten God again? Or do they out of this notion of the earth conclude, that there is no hell, or deny the punishment of sin? Do not men believe? Do they not live just, as they did before? Besides, this detracts from the dignity of your learning, and derogates from our right and title of coming to this place, that these opinions of yours may very well be true. If therefore any man have honor of title to this place in this matter, it belongs wholly to our Clavius, who opposed himself opportunely against you, and the truth, which at that time was creeping into every man's mind.

If the opinions of Copernicus "may very well be true," then his claim to a place of honor in Hell is considerably diminished. The Clavius cited above is the same Christopher Clavius mentioned in Brecht's *Galileo.* Clavius, because of his adherence to Aristotelian and Ptolemaic doctrines, had attacked Copernicus. In *Ignatius his Conclave* Donne satirizes Clavius for his reform of the Julian calendar.

In an even lighter vein Donne writes approvingly of Copernican notions when describing a convalescence in his *Devotions:*

> I am up, and I seem to stand, and I go round; and I am a new argument of the new philosophy, that the whole earth moves round; why may I not believe, that the whole earth moves in a round motion, though that seem to me to stand, when I seem to stand to my company, and yet am carried, in a giddy and circular motion, as I stand? (*Devotions upon Emergent Occasions,* 1624.)

More than in prose, Donne was able to express his inner feelings in poetry. He had many concerns—the death of Queen Elizabeth, his transition from Roman Catholic to Protestant, and his clandestine marriage, which had alienated his patrons. Each of these may have contributed in some degree to a feeling of pessimism which he voiced in "An Anatomy of the World" or "The First Anniversary" (a reference to the first anniversary of the death of the 15-year-old daughter of Sir Robert Drury, a potential patron). Changes in the world-picture wrought by Copernicus, Kepler, Brahe, and others became scientific metaphors for Donne's doubts and discontents.

> And new philosophy calls all in doubt,
> The element of fire is quite put out,
> The sun is lost, and th' earth, and no man's wit
> Can well direct him where to look for it.
> And freely men confess that this world's spent,
> When in the planets and the firmament

They seek so many new.

"Philosophy" here means "natural philosophy" or "science." In the new philosophy the sun and earth had changed places, and the earth, the stage of man, was no longer central, being merely one of many planets circling the sun. The reference to the element of fire being put out is to the sphere of fire which in Ptolemaic theory was supposed to lie between earth and moon. Kepler had decisively laid this notion to rest by suggesting an analogous situation: a crucible of live coals is placed between the eye and a wall upon which the sun is shining. The heated column of air above the coals cannot be seen directly, yet it distorts objects seen through it due to the continually varying index of refraction of the air, and its moving shadow can be seen on the wall. Stars, although they twinkle, do not appear to be undergoing a similar distortion, hence no fiery sphere exists.

In "Progress of a Soul" or "The Second Anniversary" Donne again refers to fire, but this time as one of the four elements which composed all things.

Have not all souls thought
For many ages, that our body is wrought
Of Air, and Fire, and other Elements?
And now they think of new ingredients . . .

Donne expresses doubt that fire is one of the elements, but in an earlier poem he had used the concept without question.

My fire of passion, sighs of air,
Water of tears, and earthly sad despair,
Which my materials be. ("The Dissolution")

Probably Donne had been influenced by Cardan's *De Subtilitate* (1550) in which the author denies that fire is one of the elements because it cannot nourish the body or resist decay, but can only provide heat. The mention of "new ingredients" most likely refers to the belief of Paracelsus that the body consists mainly of salt, sulphur, and mercury.

Continuing "The First Anniversary" in a spirit of melancholy the poet bemoans the passing of the old order:

'Tis all in pieces, all coherence gone;
All just supply, and all relation:
Prince, subject, father, son, are things forgot,
For every man alone thinks he hath got
To be a phoenix.

This is a reference to an ancient idea which originated with Plato and Aristotle, was maintained during the medieval period, and flourished through the late eighteenth century: the conception of a "Great Chain of Being." All orders of life were arranged in hierarchical order from the lowliest animalcules to the highest angels in steps of infinitesimal degree with man occupying a middle position. The ballad singer in the Carnival Scene (Scene 9) of Brecht's *Galileo* names the various links in the chain. It was a conception of the universe in which man could find his place. A corollary was that discord in the celestial realm had repercussions in human society, a tenet of astrology. Shakespeare, in 1602 after Copernicus had announced his geocentric theory, wrote Ulysses' speech in *Troilus and Cressida* as follows:

> The heavens themselves, the planets and this centre
> Observe degree, priority and place,
> Insisture, course, proportion, season, form,
> Office and custom, in all line of order;
> And therefore is the glorious planet Sol
> In noble eminence enthroned and sphered
> Amidst the other; whose medicinable eye
> Corrects the ill aspects of planets evil,
> And posts, like the commandment of a king,
> Sans check to good and bad: but when the planets
> In evil mixture to disorder wander,
> What plagues and what portents! What mutiny!
> . . . . .
> O, when degree is shaked,
> Which is the ladder to all high designs,
> The enterprise is sick! How could communities,
> Degrees in schools and brotherhoods in cities,
> Peaceful commerce from dividable shores,
> the primogenitive and due of birth,
> Prerogative of age, crowns, sceptres, laurels,
> But by degree, stand in authentic place?
>
> (Act I, Scene III)

The thought of Shakespeare was not demonstrably influenced by the radical notions of Copernicus. By the time of John Donne, Galileo and Kepler had produced additional evidence of the truth of the new world picture, but Donne resisted wholehearted acceptance.

The apparent irregularity of the solar orbit causes Donne some uneasiness.

> . . . Nor can the Sun
> Perfect a Circle, or maintain his way
> One inch direct; but where he rose today
> He comes no more, but with a cozening line,
> Steals by that point, and so is Serpentine.
>
> ("The First Anniversary")

As early as the second century B.C. the Greek astronomer, Hipparchus, had noticed imperfections in the solar motion. Ptolemy's followers had dealt with this motion by adding a ninth sphere containing the stars. The ninth sphere rotated once every 23 hours and 56 minutes, while the eighth sphere rotated once every 26,000 years in order to explain the precession of the equinoxes, as we now call this phenomenon. Copernicus erroneously believed that the rate of precession varied, and he struggled unsuccessfully to improve the medieval explanation by adding additional circles to his system. But followers of Copernicus elegantly accounted for the precession of the equinoxes by adding a conical motion of the earth's polar axis with a period of 26,000 years. Isaac Newton gave the currently accepted physical explanation as a gravitational tug by the moon on the earth's bulging equatorial belt.

Donne's scientific allusions are not wholly concerned with astronomy. Donne was familiar with William Gilbert's *De Magnete* (1600) and from it drew a scientific metaphor to counterbalance the world's condition of being "all in pieces, all coherence gone." In the excerpt below, "she" refers to the soul of Elizabeth Drury.

> She that should all parts to reunion bow,
> She that had all magnetic force alone,
> To draw, and fasten sundered parts in one.
>
> ("The First Anniversary")

Just as a loadstone draws together iron filings into one coherent whole, so Elizabeth Drury's soul will reunite the fragmented earth.

In "The Second Anniversary" Donne indicates his familiarity with one of the most pressing medical questions of his time.

> Know'st thou how blood, which to the heart doth flow,
> Doth from one ventricle to th' other go?
> . . . . .
> There are no passages, so that there is
> (for aught thou know'st) piercing of substances.

Galen, a second century A.D. Greek physician, had required the blood to pass right through the central wall of the heart from the right ventricle to the left. But the 16th century anatomist, Vesalius, could find "no passages." Only a few years after Donne penned these lines the English physician, William Harvey, published his book, *De Motu Cordis et Sanguinis (On the Motion of the Heart and Blood)* (1628), and brilliantly solved the riddle with his theory of the circulation of the blood.

I have been able to touch upon only a few of the scientific allusions in Donne's works. A more complete study must include an examination of his sermons and love poems as well. The new philosophy filled him not with enthusiasm but with scepticism and confusion. He skillfully used the new philosophy as a rich store of poetic imagery and symbolism, but he missed the central point of the importance of experiment and observation in establishing the new discoveries. The picture that emerges is of a man giving intellectual assent to the new philosophy while looking back longingly to the old order.

John Milton

A youthful Milton journeyed to Italy and, as he said in the *Areopagitica,* "found and visited the famous Galileo grown old, a prisoner to the Inquisition." The meeting probably took place at Galileo's house in Arcetri in 1638. What took place during the encounter between the 29-year-old Milton and the blind scientist is unknown, but Galileo may have encouraged the youth to look through his telescope. Milton's works after this bear the imprint of a mind that had become vividly aware of the infinity of the universe. Even after Milton, too, became blind, he sought out

> The secrets of the hoary deep, a dark
> Illimitable ocean without bound,
> Without dimension; where length, breadth, and height
> And time, and place are lost . . . (II, 891-894)

(Roman numerals refer to the books of *Paradise Lost,* Arabic numerals to line numbers.) The blind Milton's composition of *Paradise Lost* (1667), a feat similar to the stone-deaf Beethoven's creation of the six late quartets, evokes admiration; both are products of the mature thought of a genius working against tremendous personal odds.

In *Paradise Lost* a Protestant, moral conception of the universe is presented in phrases so mellifluous that they have been likened to the sound of an organ. Consequently, *Paradise Lost* is on nearly everyone's list of great books. Milton's conception of the temptation and fall of man is by now so much a part of Western tradition that most people confuse Milton's version with that of the Bible. For example, most people do not know that the forbidden fruit is not called an apple in the Bible, but it is so named in *Paradise Lost.* In spite of Milton's conscious attempts to portray him as the arch-villain, Satan often emerges as a hero. The stage for the conflict between the forces of good and evil is the entire universe. Marjorie Nicolson says that *Paradise Lost* is "the first modern cosmic poem in which a drama is played against a background of interstellar space." In what follows the nature of the cosmos that Milton imagined will be examined.

Milton refers to the telescope three times in *Paradise Lost.* The first is in reference to the shield of Satan which

> Hung on his shoulders like the moon, whose orb
> Through optic glass the Tuscan artist views
> At evening from the top of Fesole,
> Or in Valdarno, to descry new lands,

Rivers or mountains in her spotty globe. (I,287-291)

This passage specifically refers to Galileo and his description of mountains on the moon. Next is a reference to sunspots.

. . . a spot like which perhaps
Astronomer in the sun's lucent orb
Through his glazed optic tube yet never saw. (III,588-590)

Galileo is again mentioned in connection with the moon.

. . . As when by night the glass
Of Galileo, less assured, observes
Imagined lands and regions in the moon. (V,261-263)

In addition Milton mentions the Milky Way which Galileo had discerned to be composed of individual stars.

A broad and ample road, whose dust is gold
And pavement stars, as stars to thee appear,
Seen in the Galaxy, that milky way
Which nightly as a circling zone thou seest
Powdered with stars (VII,577-581)

The angel Uriel is said to descend

Swift as a shooting star
In autumn Thwarts the night, (IV,556-557)

an accurate observation, since modern studies of meteor showers have shown that they indeed occur more frequently in autumn. The showers take their names from the constellation from which they appear to originate.

Table 4.

Name of Shower	*Date of Maximum Display*	*Approximate Maximum Visual Hourly Count*
Lyrids	April 21	5
Eta Aquarids	May 4	5
Perseids	August 12	40
Draconids	October 10	variable
Orionids	October 21	15
Taurids	November 1	5
Andromedids	November 14	low
Leonids	November 17	10
Ursids	December 22	15

Comets are mentioned also in *Paradise Lost*. The famous Halley's Comet made an appearance in the year before Milton's birth, and he may have heard his elders speak of it. Other comets appeared in 1618 when he was a child. One reference to a comet contains a famous astronomical error.

Incensed with indignation, Satan stood
Unterrified, and like a comet burned
That fires the length of Ophiuchus huge
In the artic sky, and from his horrid hair
Shakes pestilence and war. (II,707-711)

The mistake is Milton's location of the constellation Ophiuchus in the north; actually it is closer to the ecliptic. Another reference to a comet occurs at the end of *Paradise Lost*, when Adam and Eve are led out of the Garden of Eden.

High in front advanced
The brandished sword of God before them blazed
Fierce as a comet. (XII, 632-634)

Our world is first seen through the eyes of Satan as if viewed from a great distance by means of a telescope.

And fast by hanging in a golden chain,
This pendant world, in bigness as a star
Of smallest magnitude close by the moon. (II, 1051-1053)

Milton's cosmic perspective is nowhere better displayed than in the passage in which Satan views the earth and then plunges toward it.

Satan . . .
Looks down with wonder at the sudden view
Of all this world at once . . .
Round he surveys, and well might, where he stood
So high above the circling canopy
Of night's extended shade; from eastern point
Of Libra to the fleecy star that bears
Andromeda far off Atlantic seas
Beyond the forizon; then from pole to pole
He views his breadth,—and without longer pause,
Down right into the World's first region throws
His flight precipitant, and winds with ease
Through the pure marble air his oblique way
Amongst innumerable stars that shone,
Stars distant, but nigh-hand seemed other worlds. (III, 542-566)

Milton mentions three of the four astronomical theories discussed in Chapter 2. That the earth remains stationary and the heavens rotate is rejected by Adam in his discourse with the angel Raphael on the basis of a disproportion in nature. Milton's contemporaries also commented on the fact that very excessive speeds would be required for the more distant celestial objects. Here is Adam's initial discussion of this topic.

> When I behold this goodly frame, this World,
> Of Heaven and Earth consisting, and compute
> Their magnitudes—this Earth, a spot, a grain,
> An atom, with the Firmament compared . . .
> . . . reasoning, I oft admire
> How Nature, wise and frugal, could commit
> Such disproportions, with superfluous hand
> So many nobler bodies to create
> Greater so manifold, . . .and on their Orbs impose
> Such restless revolution day by day
> Repeated, while the sedentary Earth,
> That better might with far less compass move,
> Served by more noble than herself, attains
> Her end without least motion . . . (VIII,15-35)

According to Raphael the complicated Ptolemaic system may move God to laughter.

> The great architect . . .his fabric of the heavens
> Hath left to their disputes—perhaps to move
> His laughter at their quaint opinions wide
> Hereafter, when they come to model Heaven,
> And calculate the stars; how they will wield
> The mighty frame: how build, unbuild, contrive
> To save appearances; how gird the sphere
> With centric and eccetric scribbled o'er,
> Cycle and epicycle, orb in orb. (VIII, 72-84)

The phrase, "Contrive to save appearances," epitomizes the efforts of Ptolemy and his followers to construct a geometrical model of the heavens which would correspond with observation. Although Milton sets his poem in the geocentric Ptolemaic system, his language in this passage indicates hidden disapproval. Not only does God laugh at the quaint Ptolemaic notions, but the system is so complicated that it seems to be "scribbled o'er" with too many circles. In the next passage Raphael comments on the Copernican hypothesis.

> . . . What if the sun
> Be center to the World, and other stars
> By his attractive virtue and their own
> Incited, dance about him various rounds?
> Their wandering course, now high, now low, then hid,
> Progressive, retrograde, or standing still,
> In six thou seest, and what if seventh to these
> The planet Earth so steadfast though she seem,
> Insensibly three different motions move?
> Which else to several spheres thou must ascribe,
> Moved contrary with thwart obliquities. (VIII, 122-132)

Milton finds the earth's apparent stability ("steadfast though she seem") difficult to reconcile with its triple motion in the Copernican scheme.

The triple motion of the earth mentioned by Raphael was a notion of Copernicus who thought that the earth had 1) a diurnal rotation about its own axis, 2) and annual revolution about the sun, and 3) an annual rotation about its own axis. The first motion is obvious, but the second and third motions require further explanation because they stem from a curious holdover from Aristotelian thought. Copernicus believed that the earth was embedded in a crystalline sphere similar to the one that formerly carried the earth in the heliocentric scheme, and therefore the earth's polar axis would not point to the same position on the celestial sphere during the earth's annual revolution about the sun. Thus, if the earth's polar axis were pointing away from the sun at some time of year, it would always point away from the sun, contrary to observation. Copernicus' third motion undid this by a conical motion of the earth's polar axis in a direction opposite to the annual revolution of the earth about the sun and at a rate sufficient to keep the axis of the earth parallel to itself throughout the year. Copernicus could account for the precession of the equinoxes by giving to this third motion a period of slightly less than

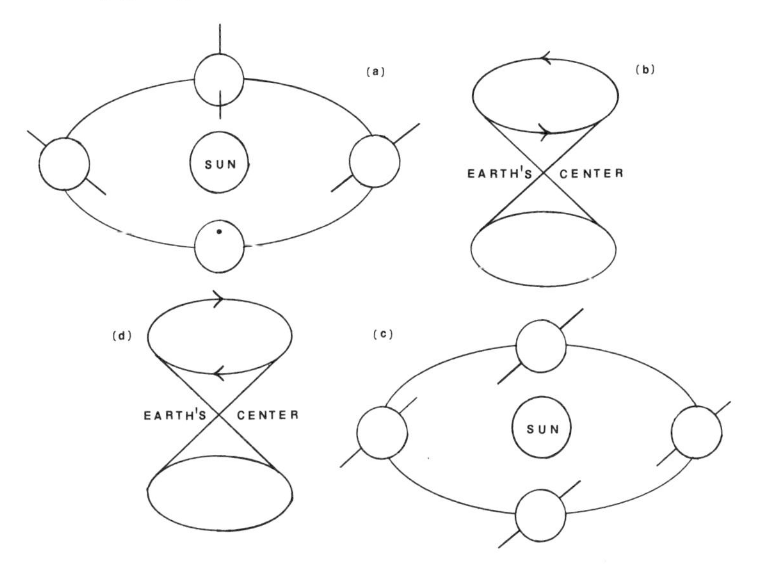

Fig. 13. (a) Copernicus' second motion of the earth; here the earth is visualized as embedded in a crystalline, heliocentric sphere. (b) The motion of the earth's axis in diagram (a) as seen by someone on the earth. (c) Actual motion of the earth with the earth's axis moving parallel to itself. (d) Copernicus' third motion necessary to convert diagram (a) to diagram (c).

one year, but his attempts to explain a fictitious change in the rate of precession were unfruitful (see previous discussion under John Donne). Kepler and Galileo did not need Copernicus' third motion because their schemes did not involve crystalline spheres which forced the earth's axis out of parallelism with itself during the year. So it would seem that Milton had not kept up with the latest scientific thinking in this area.

Milton displays little more than routine interest in the scientific astronomy which was rapidly developing in his day. He mentions the diurnal rotation of the heavens, satirizes the Ptolemaic theory, and refers to the Copernican theory without committing himself. Why doesn't he mention the Tychonian theory, the chief contemporary rival to the Copernican theory? Either he was not interested in it or thought it unimportant. Evidently a more technical analysis of these hypotheses did not suit his moral and poetical purposes in composing *Paradise Lost*.

The Influence of Sir Isaac Newton

A Scientific Profile. Newton's importance in the history of science can scarcely be exaggerated. Some idea of his scientific work is necessary in order to understand the strong poetical reaction to it.

The year 1642 marked the death of Galileo, the one-hundredth anniversary of the publication of Copernicus' *De Revolutionibus*, and the birth of Isaac Newton. Although he and Milton were contemporaries, Newton's *Philosophiae Naturalis Principia Mathematica* (1687) which brought him world fame was published twenty years after *Paradise Lost*. Thus, the full impact of Newton's discoveries is not apparent in Milton's poetry. During the years of the plague Cambridge University was closed, and Newton retired to his family home at Woolsthorpe where in his twenty-third and twenty-fourth years he conceived of a new branch of mathematics, the "Method of Fluxions," now known as differential and integral calculus. At the same time he developed the law of universal gravitation and his three laws of motion, all of which were to be published years later in the *Principia*. He said of his work of this time,

> All this was in the two plague years of 1665 and 1666, for in those days I was in the prime of my age for invention, and minded mathematics and philosophy more than at any time since.

Before the establishment of scientific societies (the Royal Society received its charter in 1662) scientific results were communicated by word of mouth or by letter. This method of communication subjected the findings to the risk of distortion or plagiarism, and authors sometimes resorted to code or gave

Fig. 14. Godfrey Kneller's portrait of Sir Isaac Newton (1702).

answers in riddles in order to protect their priority of discovery. With the advent of the Royal Society, open publication in the *Philosopohical Transactions* reduced the need for such elaborate precautions, but old habits were hard to break. Newton became involved with the German mathematician Leibniz in a controversy over who had first discovered the calculus and with Robert Hooke over gravitation and the nature of light. A committee of the Royal Society was appointed to investigate the priority of discovery of the calculus in 1712, and it decided in favor of Newton. Even this did not satisfy the continental partisans of Leibniz, because one of Newton's strong supporters, Halley, helped to write the report. Modern opinion is that the two discoveries were independent. Newton, in his public statements, shunned controversy of this sort, but when he joined battle it became a holy war of truth. Newton was a formidable adversary, pursuing his enemies to the final submission, public recantation of error. Frank E. Manuel's psychological studies of Newton reveal a man with a paranoid personality, an aspect of Newton's character that had been obscured by the 18th century near-deification of Newton. Thus, Newton made his supreme achievements in spite of psychological odds that would have severely limited lesser men.

Laplace, the French mathematician, wrote, "The *Principia* is pre-eminent above any other production of human genius." Because Newton gave the proofs in terms of geometry rather than in the calculus which he had invented to facilitate his work, the book is difficult to read today. Newton apparently did not want to confuse the soundness of his astronomical discoveries with questions about the validity of the new mathematical method. In *Principia* he included his three laws of motion. 1) His law of inertia states that everybody persists in its state of rest or of uniform motion in a straight line unless it is compelled to change that state by forces impressed on it. Newton really adopted the law of inertia which Descartes had correctly formulated in his *Principia Philosophiae* (Principles of Philosophy). 2) The second law is most simply expressed today as the familiar $F=ma$, an equation stating that force equals the product of mass times acceleration. Note that this law is consistent with the first, in that, if the force is zero, there will be no acceleration. Newton's formulation of the second law was in the form of a proportionality rather than an equation. Whiston's translation (1716) reads as follows: "The mutation of the motion is proportional to the moving Force impress'd,"or in terms of modern symbols, $\Delta(mv) \propto F$, where the symbol "Δ" means "a small change in the quantity, " and mv, the product of mass times velocity, is called momentum. 3) The content of the third law is that to every action there is always opposed an equal reaction. For example, the moon pulls the earth with the same force as the earth pulls the moon.

The *Principia* also contains his Law of Universal Gravitation, a statement that the force of gravitation is inversely proportional to the square of the distance between two interacting bodies. Robert Hooke's claim that he gave Newton the hint for this law led to a great deal of bitterness between the two. Newton later went on to prove that in the gravitational attraction between spherical bodies the mass acted as if it were concentrated at a point in the center of the sphere.

Newton then proceeded to derive Kepler's three laws of planetary motion. For uniform circular motion at least, Kepler's third law can easily be derived from Newtion's Law of Universal Gravitation. Christiaan Huyghens had found centripetal force to be proportional to v^2/r, where v is the tangential velocity, and r is the radius of the circle. Newton derived this result independently before Huyghens published it. Anyone who has tied an object on a string and swung it in a circle has applied this force by pulling on the string. Huyghens apparently never thought of applying this force to astronomical problems; his result appended a book devoted to pendulum clocks. Newton saw that the centripetal force in planetary motion is provided by gravity and is proportional to $1/r^2$. By equating the force of gravity with centripetal force it is seen that the quantity v^2r is a constant. Now all that is necessary is to replace v by its definition for circular motion, $2\pi r/T$ (circumference divided by period), and Kepler's third law, $r^3/T^2=$ constant, is obtained. Newton's completely general derivation, not restricted to circular motion, was one of the greatest triumphs of his physics.

Newton also readily worked out the motions of two bodies attracting one another, and made a good beginning on the difficult problem of three bodies under mutual gravitational attraction, for example, sun, moon, and earth. Other topics which he considered were the effect of air resistance on a pendulum, the precession of the equinoxes, and the tides. To summarize, in a sentence Newton added to the second edition, "It is enough that gravity does really exist, and act according to the laws which we have explained, and abundantly serves to account for all the motions of the celestial bodies, and of our sea."

The predictive power of the Newtonian scheme has been amply demonstrated. To give but one example, John C. Adams in England and U.J.J. Leverrier in France independently calculated that perturbations in the orbit of the planet Uranus were due to the presence of a new planet, and gave coordinates of its position. After a great deal of resistance and grumbling, astronomers in Berlin and Cambridge aimed their telescopes at the predicted portion of the sky and discovered the planet Neptune in 1846. Newtonian physics reigned supreme for about 200 years until Albert Einstein showed that Newton's laws of motion must be modified at speeds close to that of light. Einstein's General Theory of Relativity also predicts small departures from Newton's law of gravitation, but calculations based on Newtonian equations are still used because they are simpler and accurate enough for most purposes.

Newton's next book, *Opticks* (1704), published in English before a Latin translation was made in 1706, was more accessible to the non-mathematically inclined person than the formidable

Principia. Newton used little mathematics to describe his investigations of the interaction of light with mirrors, lenses, prisms, telescopes, and other familiar objects. In *Opticks* he described his construction of a reflecting telescope using a metal mirror to avoid the problems of chromatic aberration associated with the use of a glass lens. He gave an account of how he polished the one-inch diameter mirror of speculum metal, an alloy which he made himself.

Several of the experiments recorded in *Opticks* concerned the colors displayed in light which had been reflected from a thin film or refracted by a prism, sphere, or lens. In his book he recounted his experiments with prisms, how a prism dispersed a beam of white light into a colored spectrum, and how the colors could be recombined into white light by means of a second prism. In a crucial experiment he displayed the prism-generated spectrum on a board, and cut a hole in the board, allowing a spectrally-pure color to pass through it. A second prism would not split this pure

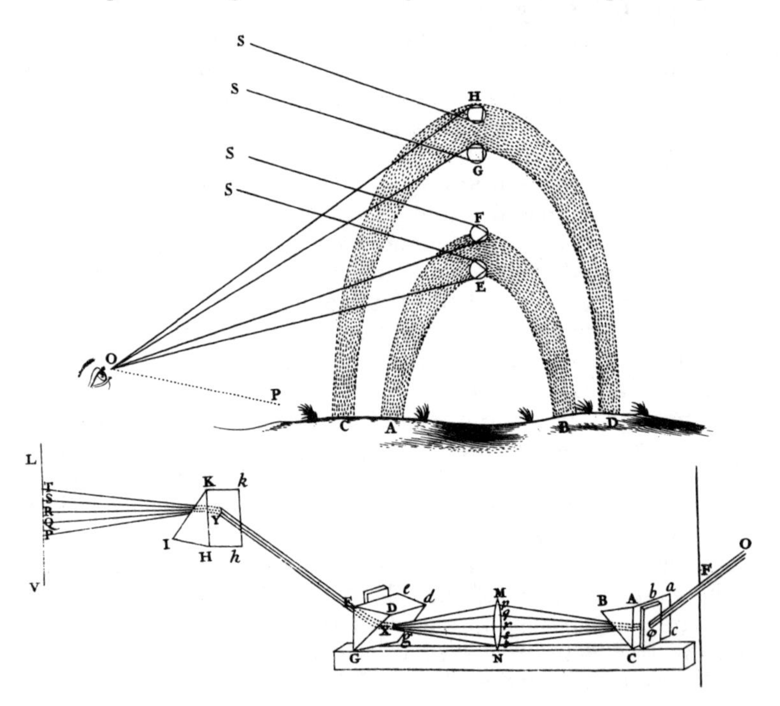

Fig. 15. Two figures from Isaac Newton's Opticks *(1704). The top illustration shows the primary and secondary rainbows. The botttom figure shows a beam of sunlight, entering from the right, being split up into a colored spectrum by the first prism, then recombined into white light by the second prism, and finally split up again by the third prism.*

color into further colors; it was bent by the second prism at exactly the same angle as the identical color was bent by the first prism. So he discovered that all colors were present in a beam of white light, and that once a particular color had been separated out of the beam, no further color change was possible. To explain the primary and secondary rainbows he invoked the "refrangibility" or refraction properties of a drop of water. He reported investigations of the colors of thin films such as soap bubbles or mica, and of the rings (now known as "Newton's rings") where a rounded surface of glass touches a flat plate. He performed an experiment on what we would call the diffraction of light (a case in which light bends around small obstacles) by using a thin slit, but failed to grasp the import of this effect.

If one's image of Newton is of a man self-absorbed and shy of controversy, then it may seem incredible that he distinguished himself as member of Parliament, Master of the Royal Mint, and President of the Royal Society. He also was deeply learned in theology, and performed experiments in chemistry and alchemy, leaving the results unpublished. The posthumous publication of his book, *The Chronology of the Ancient Kingdoms Amended,* highlighted another interest, the dating of events of ancient history.

In spite of his extraordinary achievements, Newton was very modest about his discoveries, saying,

> I do not know what I may appear to the world; but to myself I seem to have been only like a boy, playing on the seashore, and diverting myself in now and then finding a smoother pebble or a prettier shell than ordinary, while the great ocean of truth lay all undiscovered before me.

Did Newton read John Donne's sermons? Probably not, but the above passage is reminiscent of one of them. Undoubtedly, preachers of Lincolnshire and Cambridge liberally paraphrased Donne, the greatest sermonizer of his age. The relevant excerpt is as follows:

> Divers men walk by the sea-side and, the same beams of the sun giving light to them all, one gathereth by the benefit of that light pebbles or speckled shells for curious vanity, and another gathers precious pearl or medicinal amber by the same light.
>
> (Sermon preached at Saint Paul's, Christmas Day, 1621)

The statue of Newton with his prism which so impressed Wordsworth was erected in the antechapel of Trinity College, Cambridge. The simple inscription is most fitting, *"NEWTON, Qui genus humanum ingenis superavit"*(Newton, who surpassed humankind in genius).

Poetry Before and After Newton. Pre- and Post-Newtonian themes in English poetry are strikingly different. Pre-Newtonian poets, such as Donne and Milton, dealt with the twin scientific themes of the nature of the universe and man's place in it. Those questions were illuminated by the blazing light of the grand Newtonian synthesis. In the quarter-century following the death of Newton in 1727 a number of eulogies and elegies appeared in tribute to "the great Columbus of the skies," a phrase due to the poet, John Hughes. The most important of these poetical tributes was by Newton's friend, J.T. Desaguliers, who wrote in his *The Newtonian System of the World* (1729) as follows:

> Newton the unparallel'd, whose Name
> No time will wear out of the book of Fame,
> Coelestiall Science has promoted more,
> Then all the Sages that have shone before.
> Nature compell'd his piercing Mind obeys,
> And gladly shows him all her secret Ways;
> 'Gainst Mathematics she has no Defense,
> And yields t' experimental Consequence;
> His Tow'ring Genius, from its certain Cause
> Ev'ry Appearance *a priori* draws,
> And shews th' Almighty Architect's unalter'd Laws.

Such intemperate praise was gently rebuked by Alexander Pope in "An Essay on Man."

Henry Brooke, Richard Glover, Richard Blackmore, and other poets attemped to express the scientfic truths of Newtonianism directly in verse, ignoring the warning of Aristotle that poetry is not simply a metrical version of scientific statements of fact.

> The poet and historian differ not by writing in verse or in prose. The work of Herodotus might be put into verse, and it would still be a species of history, with meter no less than without it. (*Poetics* IX, 2)

In her seminal work, *Newton Demands the Muse,* Marjorie Hope Nicolson distinguishes these "scientific" poets from another group, the "descriptive" poets. The latter group took their inspiration principally from Newton's *Opticks,* in particular from his explication of light and color. "It is no exaggeration to say that Newton gave color back to poetry from which it had almost fled during the period of Cartesianism," she writes. Joseph Addison took note of this phenomenon in the *Spectator.*

> There is a second kind of beauty which . . . consists either in the gaiety and variety of colours, in the symmetry and proportion of parts, in the arrangement and disposition of bodies, or in a just mixture and concurrence of all together. Among these kinds of beauty the eye takes most delight in colours . . .For this reason we find poets, who are always addressing themselves to the imagination, borrowing more of their epithets from colours than from any other topic. (*Spectator* 412)

He might have been describing the verse of James Cawthorne,

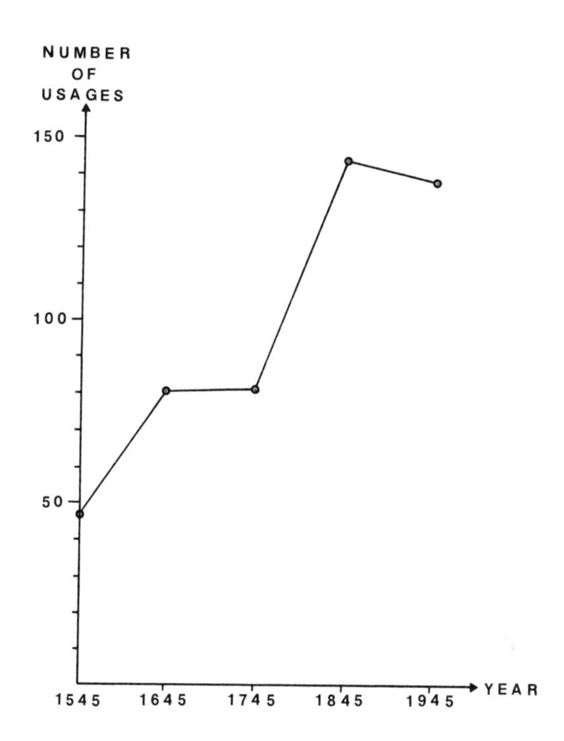

Fig. 16. The increasing occurrence of words denoting light or color in English poetry from usage tables compiled by Josephine Miles in The Continuity of Poetic Language. *"Usage" means that a word has been used 10 or more times in 1000 lines of poetry by one of a group of 20 poets writing within ±5 years of the date indicated on the abscissa.*

Mark Akenside or James Thomson.

The literary scholar, Josephine Miles, has assiduously tabulated the frequence of occurence of certain words in the English poetic vocabulary of the past 400 years. When the number of usages of words denoting light and color is graphed, one notes a striking increase as time progresses. In the period 1645-1745 several poets, most notable John Milton, used words denoting light (associated with the Godhead) and darkness (associated with Satan). the opening of Book III of *Paradise Lost* is a well-known example, "Hail holy Light, offspring of Heav'n first-born." But Newton liberated the spectrum of colors for poets, accounting for the spectacularly steep rise in number of usages between 1745 and 1845. Science continued to explore the nature of light after Newton's time. Albert Einstein's 1905 paper on the photoelectric effect (see Chapter 3) called attention to the particle-like nature of light, recalling Newton's light corpuscles. The more recent invention of the laser made available light of greater intensity, coherence (waves of light in step across an advancing wavefront) and purity of color than had ever been known before. These discoveries have widened the range of human experience, increasing the store of materials upon which poetic imagination can work.

The apotheosis of Newton ended with William Blake who died just 100 years after Newton. Newton's mathematical physics based on experimentation was not compatible with Blake's mystical vision of the world. Blake damned Newton for stifling human imagination with a "philosophy of the five senses."

> Thus the terrible race of Los and Enitharmon gave
> Laws and Religions to the Sons of Har, Binding them more
> And more to earth closing and restraining,
> Till a Philosophy of Five Senses was complete.
> Urizen wept and gave it into the hands of Newton . . .
>
> ("The Song of Los")

Many nineteenth-century poets, excluding William Wordsworth and Percy Shelley, felt that poetry was not a suitable vehicle for expounding or praising physical science. Biological science, since it had not yet been thoroughly mathematicized and was closely linked to the nature men could readily observe, was more congenial to poetic reflection. After all, Alexander Pope had urged that "The proper study of mankind is Man." So in the nineteenth century a poet, Alfred Tennyson, became a harbinger of Darwinian evolution, just as John Donne was a "barometer" of the "new philosophy" of his age. Tennyson's poem, "In Memoriam," caught the faint tremblings of thought that presaged the publication of the *Origin of Species*.

Alexander Pope

Pope had to overcome several limitations in order to pursue his career. Born a Roman Catholic in a Protestant age, he thus could not attend the usual schools. In addition he had curvature of the spine caused by tuberculosis of the bone, a fairly common disease transmitted through unpasteurized milk. His appearance was described by the painter Sir Joshua Reynolds as follows:

> He was about four feet, six inches high, very humpbacked and deformed. He wore a black coat, and according to the fashion of the time, had on a little sword. He had a large and very fine eye, and a long handsome nose; his mouth had those peculiar marks which are always found in the mouths of crooked persons; and the muscles which ran across the cheek were so strongly marked that they seemed like little cords.

Coffee-houses were especially popular in late seventeenth and early eighteenth-century London. To these establishments men would retire to smoke and imbibe the newly-introduced dark brew that dispelled sleepiness and stimulated conversation. Samuel Pepys, who later became president of the Royal Society, records in his famous diary many visits to coffee-houses where he met scientists and instrument makers of Gresham College, a name used interchangeably with the Royal Society.

> So to the coffee-house, and there fell in discourse with the Secretary of the Virtuosi of Gresham College, and had very fine discourse with him. He tells me of a new invented instrument to be tried before the College anon, and I intend to see it. (Oct. 5, 1664)

Coffee-houses have by now all but disappeared from the London scene, their places having largely been taken by exclusive men's clubs such as the Atheneum and the Garrick. Coffee-houses were an upper-class retreat, while the lower classes became addicted to gin which dulled the senses. Reaching its peak in the 1740's, the obsession for gin created social havoc, although William Hogarth's satirical etching "Gin Lane" greatly exaggerates the situation. No wonder wags, bards, and intellectuals found the sober atmosphere of the coffee-houses congenial in that hard-drinking age. A young Alexander Pope spent many happy hours in such surroundings, sharpening his wits in verbal battles with the likes of Joseph Addison and Richard Steele. In his twenty-fifth year (1713) he attended a series of popular lectures on astronomy given at Burton's coffee-house by William Whiston, a disciple and later successor of Sir Isaac Newton at Cambridge. The grandeur of the Newtonian conception tremendously impressed Pope, and after the death of Newton, when the press was filled with eulogies for the fallen giant, Pope's brief couplet stands out as the finest tribute.

> Nature, and nature's laws lay hid in night.
> God said, Let Newton be! and all was light.

One needs to add immediately Sir John Squire's famous rejoinder,

> It did not last: the Devil howling 'Ho,
> Let Einstein be,' restored the status quo.
> ("In Continuation of Pope on Newton," 1926)

The lasting impression that astronomy made on Pope's imagination may be seen most clearly in "An Essay on Man" (1733) published six years after Newton's death. (In the following, "I" and "II" refer to epistles of this poem, and Arabic numerals to line numbers.) He saw the universe as a well-ordered machine.

> But of this frame the bearings, and the ties,
> The strong connections, nice dependencies,
> Gradations just, has thy pervading soul

Looked through? or can a part contain the whole?

(I, 29-32)

The use of the words "frame" and "bearings" suggests that the poet is thinking of an orrery (Fig.48), the elaborate clockwork mechanism which showed the motions of the planets around the sun, and of the moons around the planets.

Pope epitomizes eighteenth century thought about the doctrine of "The Great Chain of Being" in several passages in this poem.

Far as creation's ample range extends,
The scale of sensual, mental powers ascends:
Mark how it mounts, to man's imperial race. (I, 207-209)

Then follows a description of animal life, culminating in this paean:

Vast chain of being! which from God began,
Natures ethereal, human, angel, man,
Beast, bird, fish, insect, what no eye can see,
No glass can reach; from Infinite to thee,
From thee to nothing. (I, 237-241)

The "just gradations" in the animate creation are reflected by similar orderings in the inanimate creation. If men were to aspire to the powers of superior beings, it would disturb the system as much as if the earth were to fly from its orbit.

. . .On superior powers
Were we to press, inferior might on outs:
Or in the full creation leave a void,
Where, one step broken, the great scale's destroyed. (I, 241-244)

Throughout the ages human beings have had difficulty coping with the spirit-matter duality. The Greeks had their temple of Dionysus built on Delos close by the temple of Apollo; later, Christians struggled with "flesh" *vs.* "Spirit." The idea of the Great Chain of Being was an underpinning of their concepts, for if there were a series of imperceptible gradations, might not the animal series terminate where the spiritual or intellectual series began? In the following famous lines Pope sees man as that intermediate link.

Placed on this isthmus of a middle state,
A being darkly wise and rudely great:
With too much knowledge for the skeptic side,
With too much weakness for the stoic's pride
He hangs between; in doubt to act or rest;
In doubt to deem himself a god, or beast;
In doubt his mind or body to prefer;
Born but to die, and reasoning but to err;

Alike in ignorance, his reason such,
Whether he thinks too little, or too much:
Chaos of thought and passion, all confused;
Still by himself abused, or disabused;
Created half to rise, and half to fall;
Great lord of all things, yet a prey to all;
Sole judge of truth, in endless error hurled:
The glory, jest, and riddle of the world! (II, 3-18)

Epistle I concludes with the words, "Whatever is, is right," an apt phrase often quoted in support of dubious conclusions. In context Pope argues that everything is part of one stupendous whole, although we may not see that whole. Local aberrations may be part of a universal order. As he says,

All chance, direction, which thou canst not see;
All discord, harmony not understood;
All partial evil, universal good. (I, 290-292)

In this spirit Pope learned to accept his deformity.

One section of the poem has often been interpreted as a criticism of Newton.

Superior beings, when of late they saw
A mortal man unfold all nature's law,
Admired such wisdom in an earthly shape,
And showed a Newton, as we show an ape. (II, 31-34)

This entire section of the poem warns against pride on the part of science. But the admirers of Newton are the ones addressed, not Newton himself. They carried praise of Newton to the point of idolatry. In this passage, "superior beings" are angelic intelligences, and Pope is reminding Newton's admirers that even a man as undeniably great as Newton in unfolding mysteries is much lower on the scale than these angels.

Pope's thesis is that, although science can accomplish wondrous things, there are definite limits to human knowledge. Science can

. . .measure earth, weigh air, and state the tides:
Instruct the planets in what orbs to run,
Correct old Time, and regulate the sun.
(II, 20-22)

Science had indeed corrected Father Time. The old Julian calendar had such inaccuracies that by the eighteenth century twelve days had been lost; some European countries had not yet adopted the calendar advocated by Pope Gregory in 1582. But questions remained which were not susceptible to scientific answers. Could science "teach Eternal wisdom how to rule?" Could man "explain his own beginning or his end?" To acknowledge our position as the middle estate in the Great Chain of Being was the proper approach:

> Know then thyself, presume not God to scan,
> The proper study of mankind is man. (II, 1-2)

William Blake

The nostalgia for the ordered world of the past noted in connection with Donne becomes in William Blake (1757-1827) a complete rejection of the scientific outlook which he associated with Newton. Blake denied that rational thought acting upon data supplied by the five senses constitutes the only true knowledge; instead he exalted the subjective powers of inspiration and imagination.

> . . .The reasoning power in man.
> This is a false body, an incrustation over my immortal
> Spirit, a selfhood which must be put off and annihilated alway.
> . . . . .
> To cast off rational demonstration by faith in the Savior;
> To cast off the rotten rags of memory by inspiration;
> To cast off Bacon, Locke and Newton from Albion's covering;
> To take off his filthy garments, and clothe him with imagination.
> ("Milton," f.41)

To Blake's eyes Bacon, Locke, and Newton constitute an unholy trinity. What had they done to deserve such censure?

Francis Bacon (1561-1626), the Lord Keeper of the Great Seal to Queen Elizabeth, sets forth his vision of a technological utopia in his book *New Atlantis*. Central to his scheme is "Salomon's House," a college of science with caves and mines for refrigeration and production of new artificial metals, towers for observation of meteorological phenomena, pools, garden houses, furnaces, perspective houses for optical experiments, and sound houses for acoustical experiments. The fellows of the college were to perform these experiments, see to practical applications of their work, and suggest more incisive experiments. Because of these speculations Bacon was honored as the spiritual father of the Royal Society, founded (1662) after his death. Bacon died as the indirect result of a scientific experiment. Seized with an idea while traveling in his carriage in winter, he was determined to test it then and there. Accordingly he purchased a fowl, and proceeded to stuff it with snow to study the effects of cold in retarding putrefaction. Unfortunately, in this episode he caught a chill which led to his death.

The philosopher John Locke (1632-1704) became a memer of the Royal Society five years after its founding, and thus was able to keep up with the latest developments in science. In addition he was a friend of Robert Boyle, one of the founders of modern chemistry. His most famous book, *An Essay Concerning Human*

Fig. 17. Frontispiece of Thomas Sprat's History of the Royal Society *(1667). From left to right, Lord Brouncker (the first president of the Society), a bust of Charles II, and Francis Bacon.*

Understanding (1690), became almost a humanist bible in some quarters, but Oxford students were forbidden to read it in 1703 because it was thought to give rise to skepticism. Locke's purpose in writing it was "to inquire into the origin, certainty, and extent of human knowledge, together with the grounds and degrees of belief, opinion and assent." He argues against the "innateness" of any part of man's knowledge. Knowledge consists of ideas or meanings referring to the data of the five senses or to operations of the human mind in reflecting upon sense data.

To be meaningful, words must therefore refer to the sensuous or mental. Deductive or inductive reasoning then leads to more general conclusions about the world.

Sir Isaac Newton's successful application of mathematical physics to the solution of long-standing problems in astronomy led to his virtual deification in England. In reaction, Blake wrote of the deadening influence of Bacon, Locke, and Newton in cartoon-like images.

For Bacon and Newton, sheathed in dismal steel, their terrors hang
Like iron scourges over Albion; reasonings like vast serpents
Enfold around my limbs, bruising my minute articulations.
I turn my eyes to the schools & universities of Europe
And there behold the loom of Locke whose woof rages dire,
Washed by the water-wheels of Newton. Black the cloth
In heavy wreaths folds over every nation; cruel works
Of many wheels I view, wheel without wheel, with cogs tyrannic
Moving by compulsion each other: not as those in Eden, which
Wheel within wheel in freedom revolve, in harmony & peace.

("Jerusalem," f.15)

In his revulsion both from the mechanistic scheme of the universe and the industrial revolution, Blake turned inward. He reveals his poetic purpose in the following:

. . . I rest not from my great task—
To open eternal worlds, to open the immortal eyes
Of man inwards into the worlds of thought—into Eternity
Ever expanding in the bosom of God, the human imagination.

("Jerusalem," f.5)

Obvious risks accompany the pursuit of these lofty ideals; a poet may have such personal visions that they cannot be adequately expressed in words, much less effectively communicated to other people. These considerations and the lack of recognition of his poems and engravings did not deter Blake. His muse would not let him rest; in the words of his character Los,

'I must create a system, or be enslaved by another man's;
I will not reason and compare; my business is to create.'

("Jerusalem," f.10)

Whether or not one agrees with Blake that the truth of poetic insight transcends the truth of science, one has to admire his heroic attempt to make poetry perform a difficult task.

William Wordsworth

If William Wordsworth's poetry is ambiguous regarding science, it reflects the ambiguity of an age that hailed the triumphs of pure science but deplored the ungracious materialism of the industrial revolution. Those who see disdain for science in his poetry base this view primarily upon two poems. One, "A Poet's

Epitaph," composed in 1799, has been the subject of much discussion.

> Art thou a Statist in the van
> Of public conflicts trained and bred?
> —First learn to love one living man;
> *Then* may'st thou think upon the dead.
> . . . . .
> Physician art thou?—one, all eyes,
> Philosopher!—a fingering slave,
> One that would peep and botanize
> Upon his mother's grave?

("Statist" is used in the sense of "statesman"; "fingering" in the sense of "stealing.") Most people would agree that the scientist has picked an inappropriate place for botanizing. Taken in context, the lines venerate individual experience, and deplore abstract categorization whether by politician, lawyer, priest, soldier, physician, or moralist. Science as such is not singled out for rebuke. Wordsworth reserves his disdain for book-learning as opposed to experience.

> One impulse from a vernal wood
> May teach you more of man,
> Of moral evil and of good,
> Than all the sages can.
>
> Sweet is the lore which Nature brings;
> Our middling intellect
> Mis-shapes the beauteous forms of things:—
> We murder to dissect. ("The Tables Turned" 1798)

This poem stresses the virtue of receiving over seeking, and synthesis over analysis.

But Wordsworth's view of science was not static. From 1800 on he was in close touch with Humphrey Davy, already a promising young chemist, and Sir William Rowan Hamilton, the Royal Astronomer for Ireland. These contacts with distinguished men of science undoubtedly influenced his thought as revealed in the "Preface to the Second Edition . . .of Lyrical Ballads."

> The poet . . .will be ready to follow the steps of the Man of science, not only in those general indirect effects, but he will be at his side, carrying sensation into the midst of the objects of the science itself. The remotest discoveries of the Chemist, the Botanist, or Mineralogist, will be as proper objects of the Poet's art as any upon which it can be employed . . .

Coming at a time when England was in the throes of the industrial revolution, this surprising and brave prophecy has only been imperfectly realized.

Wordsworth's greatest poem, "The Prelude," is addressed to

his friend Samuel Taylor Coleridge, a poet interested in science. The poem is autobiographical, and Book III describes his residence at Trinity College, Cambridge, where his bedroom window looked out on the chapel.

> And from my pillow, looking forth by light
> Of moon or favoring stars, I could behold
> The antechapel where the statue stood
> Of Newton with his prism and silent face,
> The marble index of a mind for ever
> Voyaging through strange seas of Thought, alone. (III, 58-63)

In Book V Wordsworth records a dream which followed a reading of Cervantes. De Quincy has stated that it "reaches the very *ne plus ultra* of sublimity, expressly framed to illustrate the eternity . . . of those two hemispheres, as it were, that compose the total world of human power, mathematics on the one hand, and poetry on the other." In the dream Wordsworth encounters an Arab mounted on a camel and carrying a stone (science) and a shell (the arts).

> . . . the Arab told me that the stone
> (to give it in the language of the dream)
> Was 'Euclid's Elements'; and 'This,' said he,
> 'Is something of more worth,' and at the word
> Stretched forth the shell, so beautiful in shape,
> In colour so resplendent, with command
> That I should hold it to my ear.

Wordsworth intimates that there are different roles for science and the arts.

> The one that held acquaintance with the stars
> And wedded soul to soul in purest bond
> Of reason, undisturbed by space or time;
> The other that was a god, yea many gods,
> Had voices more than all the winds, with power
> To exhilarate the spirit, and to soothe,
> Through every clime, the heart of human kind. (V, 56-140)

In Book VI follows, according to B. Ifor Evans, "one of the most elaborate and closely argued passages on science in English poetry." The term "geometric science" refers to astronomy, which was taught at Cambridge as a form of applied geometry. The passage is worth quoting at some length.

> Yet may we not entirely overlook
> The pleasure gathered from the rudiments
> Of geometric science. Though advanced
> In these inquiries, with regret I speak,
> No farther than the threshold, there I found
> Both elevation and composed delight:

With Indian awe and wonder, ignorance pleased
With its own struggles, did I meditate
On the relation those abstractions bear
To Nature's laws, and by what process led,
Those immaterial agents bowed their heads
Duly to serve the mind of earth-born man;
From star to star, from kindred sphere to sphere,
From system to system without end.

More frequently from the same source I drew
A pleasure quiet and profound, a sense
Of permanent and universal sway,
And paramount belief; there, recognized
A type, for finite natures, of the one
Supreme Existence, the surpassing life
Which—to the boundaries of space and time,
Superior, and incapable of change,
Nor touched by the welterings of passion—is,
And hath the name of, God. Transcendent peace
And silence did await upon these thoughts
That were a frequent comfort to my youth.

Wordsworth then tells of a shipwreck victim who has brought along one book, a treatise on geometry.

. . . Mighty is the charm
Of those abstractions to a mind beset
With images, and haunted by herself,
And specially delightful unto me
Was that clear synthesis built up aloft
So gracefully; even then when it appeared
Not more than a mere plaything, or a toy
To sense embodied: not the thing it is
In verity, an independent world,
Created out of pure intelligence. (VI,115-167)

As Wordsworth grew older his Christian belief became increasingly more orthodox, and his qualms concerning science increased. He was speaking for an England in the grip of the industrial revolution when he said.

True is it Nature hides
her treasures less and less. —Man now presides
In power, where once he trembled in his weakness;
Science advances with gigantic strides;
But are we aught enriched in love and meekness?
("To the Planet Venus," 1838)

Alfred Lord Tennyson

The English poet most knowledgeable and most interested in science in the nineteenth century was Alfred Tennyson (1809-1892). He was a friend of Sir Norman Lockyer, one of the leading

Fig. 18. The Great Nebula in Orion *photographed through the 100-inch telescope on Mount Wilson.*

British astronomers of that time. When Lockyer came to visit, they used the 2-inch telescope kept at Tennyson's home, Aldworth. Because he was an amateur astronomer, Tennyson's descriptions of the heavens are particularly accurate. He wrote the following lines in 1833:

> She saw the snowy poles of moonless Mars,
> That marvellous round of milky light
> Below Orion, and those double stars

Whereof the one more bright
Is circled by the other ("The Palace of Art")

But in later years he realized that he had located the nebula wrongly, below rather than in Orion, and in 1877 the two satellites of Mars predicted by Kepler were discovered. He therefore revised the poem so that it now reads as follows:

She saw the snowy poles and moons of Mars,
That marvellous field of drifted light
In mid Orion, and the married stars—

In "Locksley Hall" he gives another description of Orion, and of the Pleiades, which most certainly came from personal observation,

Many a night I saw the Pleiads, rising thro' the mellow shade,
Glitter like a swarm of fireflies tangled in a silver braid.

a good description of the Pleiades star cluster, since many of its members are surrounded with faint nebulosities.

"In Memoriam" (1850), published nine years before Darwin's *Origin if Species,* is in harmony with the best scientific thinking of the time.

Are God and Nature then at strife,
That Nature lends such evil dreams?
So careful of the type she seems,
So careless of the single life,
That I, considering everywhere
Her secret meaning in her deeds,
And finding that of fifty seeds
She often brings but one to bear,
. . . . .
'So careful of the type?' but no.
From scarped sliff and quarried stone
She cries, 'A thousand types are gone;
I care for nothing, all shall go.'
. . . . .
Who trusted God was love indeed
And love Creation's final law—
Tho' Nature, red in tooth and claw
With ravine, shrieked against his creed— (LV,LVI)

Tennyson's problem is to reconcile his belief in a benevolent and just Deity with nature as described above. The answer lies "behind the veil" (LVI). After finding his scientific observations unimpeachable, many modern readers balk at accepting his solution to the theological questions raised in connection with evolutionary theory. In a crucial episode in the poem (XCV), Tennyson enters a mystical trance and encounters the spirit of the dead Arthur Hallam to whom the poem is dedicated. This was not unacceptable to his contemporaries. When Prince Consort Albert died, Queen Victoria said, "Next to the Bible 'In Memoriam' is my comfort." In the final section of the poem, Tennyson again refers to the trance and writes of meeting one "of those that, eye to eye, shall look on Knowledge." To them "Nature [is] like an open book." He had been assured that the "whole creation moves" toward "one far-off divine event" (CXXXI).

Conclusion

Poetry and science must interact because they have a common starting point in the world of nature, if "nature" is construed to include not only facts, but also human feelings and intuitions, as well as constructs such as atoms and nuclei. This chapter has only touched the surface of these complex interactions. We have seen how poetic themes have changed according to the prevailing temper of scientific thought. Before the time of Newton speculation centered on the nature of the universe and man's place in the scheme of things; after Newton poets concentrated on this world, scrutinizing man's self-image and human knowledge in the light of the Newtonian advances. Sometimes scientists' personal acquaintance with poets affected their verse as in Kepler's influence on Donne, Galileo's on Milton, Whiston's on Pope, and Lockyer's on Tennyson. The obscure printmaker, William Blake, may never have met a famous scientist, but he reacted strongly against scientific rationalism. The same perceived challenges from science evoked similar responses from both poets and science-fiction authors (see Chap. 2). Compare Swift's trenchant criticisms of science with Pope's measured judgements in "An Essay on Man," and similarly compare Hawthorne's concern about inappropriate biological experiments with Tennyson's searching questions regarding evolution. Wordsworth's definition of poetry includes science; I prefer to let him have the last word in this discussion of poetic-scientific interactions:

> Poetry is the breath and finer spirit of all knowledge; it is the impassioned expression which is in the countenance of all Science.
>
> ("Preface to Lyrical Ballads")

Bibliography

Andrade, E.N. da C., *Sir Isaac Newton* (Anchor Books, Garden City, 1954).

Anthony, H.D., *Sir Isaac Newton* (Abelard-Shuman, London, 1960).

Applebaum, Wilbur, "Donne's Meeting with Kepler: A Previously Unknown Episode," Philological Quarterly, L, 132, January, 1971.

Ault, Donald D., *Visionary Physics: Blake's Response to Newton* (University of Chicago Press, 1974).

Bell, Arthur E., *Newtonian Science* (Edward Arnold, London, 1961).

Bronowski, J., *William Blake and the Age of Revolution* (Routledge and Kegan Paul, London, 1965).
Buchanan, Scott, *Poetry and Mathematics* (University of Chicago Press, 1962).
Bush, Douglas, *Science and English Poetry* (Oxford university Press, N.Y., 1950).
_____. "Science and Literature," in *Seventeenth Century Science and the Arts*, Rhys, H.H., Ed. (Princeton Univ. Press, 1961).
Byard, Margaret M., "Poetic Responses to the Copernican Revolution," Sci. Am. **236**, 121, June 1977.
Cadden, John J., and Brostowin, Patrick R., *Science and Literature: A Reader* (D.C. Heath and Co., Boston, 1964).
Coffin, Charles Monroe, *John Donne and the New Philosophy* (The Humanities Press, N.Y., 1958).
Crum, Ralph B., *Scientific Thought in Poetry* (Columbia Univ. Press, N.Y., 1931).
Curry, Walter C., *Chaucer and the Medieval Sciences* (Barnes and Noble, Inc., N.Y., 1960).
Davenport, William H., "Resource Letter TLA-1 on Technology, Literature, and Art since World War II," Am. J. Physics, **38**, 407, 1970.
De Selincourt, Ernest, "The Interplay of Literature and Science During the Last Three Centuries", in *Wordsworthian and Other Studies* (Russell and Russell, New York, 1964).
Dudley, F.A., *The Relations of Literature and Science: A Selected Bibliography 1930-1967* (University Microfilms, Ann Arbor, 1968).
Eastwood, W., *A Book of Science Verse* (Macmillan and Co. Ltd., London, 1961).
Gilbert, Allan H., "Milton and Galileo," Studies in Philology **19**, 152, 1922.
Grabo, Carl, *A Newton Among Poets: Shelley's use of Science in Prometheus Unbound* (University of N. Carolina Press, Chapel Hill, 1930).
Hall, A. Rupert, "Newton in France: a New View," Hist. of Science **13**, 233, 1975.
Huxley, Aldous, *Literature and Science* (Harper and Row, N.Y., 1963).
Johnson, Francis R., *Astronomical Thought in Renaissance England: A Study of English Scientific Writing from 1500 to 1645* (The Johns Hopkins Press, Baltimore, 1937).
Jones, William P., *The Rhetoric of Science: A Study of Scientific Ideas and Imagery in Eighteenth-Century English Poetry* (Univ. of California Press, Berkeley, 1966).
Keynes, Geoffrey, *A Bibliography of Dr. John Donne*, 4th edition, (Oxford University Press, 1973).
Lewis, C.S., *The Discarded Image: An Introduction to Medieval and Renaissance Literature* (Cambridge Univ. Press, 1964).
Lovejoy, Arthur O., *The Great Chain of Being: A Study of the History of an Idea* (Harvard Univ. Press, Cambridge, 1942).
Manuel, Frank E., *A Portrait of Isaac Newton* (Harvard Univ. Press, Cambridge, 1968).
McColley, Grant, *Literature and Science: An Anthology from English and American Literature, 1600-1900* (Packard and Co., Chicago, 1940).

____. "The Astronomy of *Paradise Lost*," Studies in Philology **34**, 209, 1937.

Meadows, A.J., *The High Firmament* (Leicester University Press, 1969).

Miles, Josephine, *The Continuity of Poetic Language: The Primary Language of Poetry, 1540's-1940's* (Octagon Books, New York, 1965).

Nicolson, Marjorie Hope, *The Breaking of the Circle: Studies in the Effect of the "New Science" Upon Seventeenth Century Poetry* (Northwestern Univ. Press, Evanston, 1950).

____. *Newton Demands the Muse: Newton's "Opticks" and the Eighteenth Century Poets* (Archon Books, Hamden, Conn., 1963).

____. *Mountain Gloom and Mountain Glory: The Development of the Aesthetics of the Infinite* (Cornell Univ. Press, Ithaca, 1959).

____. *Pepys' "Diary" and the New Science* (Univ. Press of Virginia, Charlottesville, 1965).

____. "Resource Letter SL-1 on Science and Literature," Am.J. Phys. **33**, 175, 1965.

____. *Science and Imagination* (Great Seal Books, Ithaca, 1956).

____. *Voyages to the Moon* (Macmillan Co., N.Y., 1948).

Nicolson, Marjorie Hope, and Rousseau, G.S., *"This Long Disease, My Life": Alexander Pope and the Sciences* (Princeton Univ. Press, 1968).

Svendsen, Kester, *Milton and Science* (Harvard Univ. Press, Cambridge, 1956).

Chapter 5

Science in American Poetry

THE STYLE OF American poetry was at first only an echo of the style of British poetry. Nevertheless, new forces were at work: the frontier was being pushed westward, cities were developing, traditional religious belief was being challenged by the transcendentalism of Emerson and Thoreau, and, in the nineteenth century, democratic ideals were put to the supreme test in the Civil War. Out of this turbulent new democracy emerged a poet who spoke with an authentic American voice, Walt Whitman. Deliberately purging his vocabulary of stock poetic phrases, he developed what he called a "recitative." Although Tennyson and Whitman died in the same year (1892), their poetical expressions were so different that they seem to be separated by generations. Tennyson agonized over the conflict he saw between the material and the spiritual; Whitman felt that the material and the spiritual were one, and that evolution was a means toward divinity. After Whitman I shall consider a New England poet, and two poets writing from the vantage point of the West Coast. Finally, as a reminder that the ingredients of the melting pot came not only from Great Britain but from other countries, I shall examine the poetry of an Italian-American, John Ciardi.

Walt Whitman

In the poem "Song of Myself" (1855) Whitman gives the fullest statement of his evolutionary beliefs. That "barbaric yawp," as he called it, is indeed self-revealing, but one must not expect a deep self-analysis, because he said, "I have no chair, no church, no philosophy" (Stanza 46). Although he celebrated science for its usefulness, he was not deeply schooled in it.

> I accept Reality and dare not question it,
> Materialism first and last imbuing.
>
> Hurrah for positive science! long live exact demonstration!
> Fetch stonecrop mixt with cedar and branches of lilac,

This is the lexicographer, this the chemist, this made a grammar of old cartouches,
These mariners put the ship through dangerous unknown seas,
This is the geologist, this works with the scalpel, and this is a mathematician.

Gentlemen, to you first honors always!
Your facts are useful, and yet they are not my dwelling,
I but enter by them to an area of my dwelling. (23)

Like Wordsworth he preferred direct experience to book-learning.

When I heard the learn'd astronomer,
. . . . .
How soon unaccountable I became tired and sick,
Till rising and gliding out I wander'd off by myself,
In the mystical moist night air, and from time to time,
Look'd up in perfect silence at the stars.
("When I Heard the Learn'd Astronomer")

He never doubted that man was making progress in his evolution to a higher destiny, and that the process could not be stopped.

There is no stoppage and never can be stoppage,
If I, you, and the worlds, and all beneath or upon their surfaces, were this moment reduced back to a pallid float, it would not avail in the long run,
We should surely bring up again where we now stand,
And surely go as much farther, and then farther and farther.

He follows this passage with an attempt to describe the infinity of space and time.

A few quadrillions of eras, a few octillions of cubic leagues, do not hazard the span or make it impatient,
They are but parts, any thing is but a part.
See ever so far, there is limitless space outside of that,
Count ever so much, there is limitless time around that.
("Song of Myself," 45)

For Whitman soul and body are often indistinguishable; "Behold, the body includes and is the meaning, the main concern, and includes and is the soul" ("Starting from Pauanok," 13). He describes the journey of the everlasting soul up the steps of evolution in the passage below from "Song of Myself."

My feet strike an apex of the apices of the stairs,
On every step bunches of ages, and larger bunches between the steps,
All below duly travel'd, and still I mount and mount.

Rise after rise bow the phantoms behind me,
Afar down I see the huge first Nothing, I know I was even there,
I waited unseen and always, and slept through the lethargic mist,
And took my time, and took no hurt from the fetid carbon.

. . . . .
My embryo has never been torpid, nothing could overlay it.
For it the nebula cohered to an orb,
The long slow strata piled to rest it on,
Vast vegetables gave it sustenance,
Monstrous sauroids transported it in their mouths and deposited it
with care. (44)

Death will reduce his body to dust, grist for the mill of evolution.

I bequeath myself to the dirt to grow from the grass I love,
If you want me again look for me under your boot-soles. (52)

This reduction to atoms is the ultimate basis for his passionate belief in democracy.

I celebrate myself, and sing myself,
And what I assume you shall assume,
For every atom belonging to me as good belongs to you. (1)

The same thought occurs in "On the Beach at Night Alone." "A vast similitude interlocks all," he says, whether they be suns, moons, planets, souls, or animals.

Like Tennyson, Whitman would revise a poem if an astronomical reference proved incorrect. In the 1871 edition of "On the Beach at Night" he referred to the Pleiades as "brothers," but in Greek mythology the Pleiades are the seven *daughters* of Atlas. Only six stars are visible to the naked eye (through a telescope hundreds more stars are visible), implying that one daughter refuses to show herself out of shame for having loved a mortal. Realizing his boner, Whitman changed "brothers" to "sisters" in the 1876 edition.

In the Preface to the 1855 edition of *Leaves of Grass* Whitman described his views on the relation of poetry to science.

> The sailor and traveler . . . The atomist chemist astronomer geologist phrenologist spiritualist mathematician historian and lexicographer are not poets, but they are the lawgivers of poets and their construction underlies the structure of every perfect poem. No matter what rises or is uttered they sent the seed of the conception of it . . . of them and by them stand the visible proofs of souls . . . always of their fatherstuff must be begotten the sinewy races of bards. If there shall be love and content between the father and the son and if the greatness of the son is the exuding of the greatness of the father there shall be love between the poet and the man of demonstrable science. In the beauty of poems are the tuft and final applause of science.

(Here "tuft" refers to the gold tassel worn by titled students at English universities.) In prose as well as poetry Whitman was plain-spoken. The above passage lacks the stylistic elegance of Wordsworth's "Preface to Lyrical Ballads" written a half-century earlier (see Chap. 4), yet each in its own way affirms the curious

Fig. 19. The Pleiades cluster showing nebulosity.

affinity of science and poetry. Each poet calls the roll of the scientists' occupations: chemist, mathematician, etc. Whitman characteristically adds an element of masculine love; he compares the scientist-poet relationship with the father-son relationship. Whitman would be proud to be remembered as a great lover whose love knew no bounds. In an earlier passage in the Preface he extends his love to cover the entire physical universe, "The known universe has one complete lover and that is the greatest poet."

Robert Frost

The process of simplifying the language of poetry begun by Walt Whitman continues with Robert Frost (1874–1963). Although Frost speaks plainly, he does not usually speak disturbingly; Whitman's ferocity of expression, homosexuality, and praise of the body, shocking to the naive reader, are absent here. In spite of the country's harrowing experiences in the Second World War, Frost rarely writes of the war or the Bomb, an exception being the poem, "U.S. 1946 King's X." In contrast, Whitman writes movingly of his experiences as a war correspondent and nurse for wounded soldiers during the Civil War. But all is not blandness in Frost's work, because he occasionally draws an image of terror from nature as in his poem, "Design," where the triple whitenesses of flower, spider, and its victim seem more than coincidental. Frost became the most popular American poet of his age, and in 1961 he was invited to read one of his poems at the Inaugural of President John F. Kennedy, an official recognition as close to Poet Laureate as a country lacking such an office can award.

I like to think of Frost as a "literate farmer," a character in one of his poems. Many of his images are those that might occur to a farmer-naturalist: "Northern Lights that run like tingling nerves" ("On Looking Up by Chance at the Constellations"), "And a masked moon had spread down compass rays" ("Moon Compasses"), "The great Overdog / That heavenly beast" ("Canis Major"), "Where showers of charted meteors let fly" ("A Star in a Stoneboat"), "You know Orion always comes up sideways. / Throwing a leg up over our fence of mountains" ("The Star-Splitter"). Occasionally he drops the mask of rustic simplicity and more complicated images emerge. In "The Bear" Frost begins ordinarily enough with a description of a bear in the woods, and then he shifts the meaning to man's relation to the universe.

> The world has room to make a bear feel free;
> The universe seems cramped to you and me.

Man's position is more like that of a caged bear pacing back and forth.

> The telescope at one end of his beat,
> And at the other end the microscope,
> Two instruments of nearly equal hope,
> And in conjunction giving quite a spread.
> . . . . .
> At one extreme agreeing with one Greek,
> At the other agreeing with another Greek.

The poet seems to be asking, "What is man's place in the cosmos?" The two Greeks may refer to Democritus with his idea of atoms, and to Ptolemy with his celestial model. Frost's placing of scientific instruments at each end of a bear's cage may seem incongruous, but he does it intentionally to lighten the poem's mood.

Frost sometimes chooses unusual scientific metaphors. Helium is used to denote religious faith in "Innate Helium." The "Millikan mote" in "A Wish to Comply" refers to the oil-droplets in the Nobel Prize-winning experiment performed by Robert Andrews Millikan, the American physicist. By observing the motion of tiny oil-droplets he established that they bore electric charges in discrete units, the smallest being the charge of just one electron. The astronomical event described in "An Unstamped Letter in Our Rural Letter Box" is highly improbable, however poetic the image may seem.

> The largest firedrop ever formed
> From two stars' having coalesced
> Went streaking molten down the west.

The accuracy of the first two lines from "Skeptic" can also be questioned.

> Far star that tickles for me my sensitive plate
> And fries a couple of ebon atoms white.

The silver atoms on a developed photographic plate (i.e., a negative) form black lumps where light has hit, not white. The high-flown "ebon," in conjunction with the colloquial "fries a couple . . .," is another exmaple of Frost's deliberate humor.

In "West-Running Brook," one of Frost's finest poems, a facet of nature, keenly observed, is elevated to profound significance: a reflex water wave signifies man's fight against the "universal cataract of death."

> The black steam, catching on a sunken rock,
> Flung backward on itself in one white wave
> And the white water rode the black forever.

To a scientist, this passage conjures up rich associations with the central concepts of thermodynamics. Frost's metaphor of the wave flung backward sums up these concepts beautifully and concisely.

The "black stream" suggests the cycle of evaporation, transportation, and condensation of water vapor in the form of rain. If it were not for the sun's energizing this cycle, all water on earth would eventually find its way to one universal ocean, and rivers would cease to run. No flowing water means no possibility of transformation or change. Such a state of disorder is possible, because the sun itself is slowly using up its store of nuclear fuel; it cannot burn forever. In 1865 Rudolph Clausius summed up the situation as follows:

> We can express the fundamental laws of the universe which correspond to the two fundamental laws of the mechanical theory of heat in the following simple form.
> 1. The energy of the universe is constant.
> 2. The entropy of the universe tends toward a maximum.

These statements are known as the First and Second Laws of Thermodynamics. The term, "entropy," coming from a Greek word meaning transformation, is a measure of the amount of disorder in the universe. As the sun burns out and water seeks a uniform level, energy becomes less and less available, and the world ends in a "heat death,"[6] a prediction of the Second Law of Thermodynamics.

But human beings can locally decrease entropy by creating order out of disorder. To paint a picture or compose music is to create more order in the universe. Does this mean that the Second Law of Thermodynamics has been defied? No, because the energy making possible the act of painting or composition came from calories in food ultimately derived from the sun. Entropy in the universe *as a whole* increases in spite of the local decrease resulting from human effort. The reflex water wave as well as the title of the poem (the brook runs counter to the usual direction) imply that human beings perform a Sisyphean task; for a little while in a restricted area they heroically oppose the Second Law of Thermodynamics which inexorably leads life and substance to death and decay.

> It is from this in nature we are from.
> It is most us.

Robinson Jeffers

Not since Tennyson has there emerged a poet as thoroughly imbued with science as Robinson Jeffers (1867–1962). Jeffers' education as a graduate student began with literature, branched to medicine at U.S.C., and ended with forestry and zoology at the University of Washington. His brother, Hamilton Jeffers, was a professional astronomer, and perhaps because of this Robinson Jeffers maintained a lifelong interest in astronomy, keeping up with the latest discoveries and visiting observatories.

Many astronomical references occur in the preface to "Margrave" (1931), a narrative poem of moderate length which explores the pathologic personality of a man who kidnaps and murders a child.

[6]H.G. Wells used this idea in his novel, *The Time Machine* (1895), where in Chapter 11 the Time Traveller finds the earth of 30 million years hence resting with one face permanently towards the sun, a bloated, red hulk.

Fig. 20. Robinson Jeffers in 1941 on the occasion of his visit to lecture at the Library of Congress.

> The earth was the world and man was its measure, but our minds have looked
> Through the little mock-dome of heaven the telescope-slotted observatory eyeball, there space and multitude came in
> And the earth is a particle of dust by a sand-grain sun, lost in a nameless cove of the shores of a continent.
> Galaxy on Galaxy, innumerable swirls of innumerable stars, endured as it were forever and humanity
> Came into being, its two or three million years are a moment, in a moment it will certainly cease out from being
> And galaxy on galaxy endure after that as it were forever . . .

In contrast, William Blake aimed

> to see the World in a grain of sand
> And a heaven in a wild flower. ("Auguries of Innocence")

Here Jeffers sees the sun *as* a grain of sand, and the earth a mere particle of dust in comparison. Blake, in his mystic vision, sought transcendent truth in the commonplace; conversely Jeffers, in his scientific simile, reduces cosmic sizes to the comprehensible. "Shores of a continent" may refer both to the North American nebula and to Jeffers' home on the Big Sur coast. In the post-Copernican perspective young Walter Margrave's crime is insignificant, a thought which reduces man's role to the vanishing point.

Fig. 21. The North American nebula (NGC 7000) in the constellation Cygnus.

Jeffers' poem continues:

But man is conscious,
He brings the world to focus in a feeling brain,
In a net of nerves catches the splendor of things,
Breaks the somnambulism of nature . . . His distinction perhaps,
Hardly his advantage. To slaver for contemptible pleasures
And scream with pain, are hardly an advantage.
Consciousness? The learned astronomer
Analyzing the light of most remote star-swirls
has found them—or a trick of distance deludes his prism—

All at incredible speeds fleeing outward from ours.
I thought, no doubt they are fleeing the contagion
Of consciousness that infects this corner of space.

Jeffers' "learned astronomer" echoes the title of Whitman's 1865 poem, "When I Heard the Learn'd Astronomer," but Jeffers' reference has additional connotations, since many astronomical discoveries had since been made. Whitman could have been aware of the Doppler effect, the variation of the pitch of sound with the motion of its source (or of the observer). When applied to astronomy, the Doppler effect predicts that light from a star or galaxy receding from the earth will be reddened (i.e., lowered in frequency). In 1868 the motions of stars were first measured using this effect by passing starlight (and later the dimmer light from galaxies—Jeffers' "star-swirls") through a prism or other optical element to disperse it into its component colors. By doing this, astronomers discovered that all stars are receding from earth at tremendous speeds. As usual, Jeffers is completely accurate in his scientific reference; the galaxies are indeed "fleeing outward from ours" at "incredible speeds."

In 1929, two years before Jeffers composed this poem, Edwin Hubble formulated his Law of Recession stating that the farther away a galaxy is from earth, the faster its velocity of recession. Hubble's Law was not universally accepted, and Jeffers alludes to alternative interpretations of the astronomical data by saying, "or a trick of distance deludes his prism."

Unfortunately for traditionalists, Hubble's Law of Recession does not help to restore earth to its honored place at the center of the universe. To see why, imagine a balloon with a polka-dotted surface. As the balloon is inflated, each dot moves farther away from all other dots, but this does not make one particular dot central or unique. If the dots were galaxies, and the balloon the universe, the analogy would explain why most objects appear to be rushing away from the earth. The balloon analogy should not be taken too seriously; for one thing, the galaxies themselves do not expand as do the dots on the balloon's surface.

Jeffers' poem gives evidence of his familiarity not only with astronomy but also with biological chemistry. Near the end of "Margrave" the kidnapper is hanged. The poet theorizes that the electrochemical activity of the brain must continue for a few minutes after death. During this time the brain continues to function disconnected from the outside world much as in sleep or schizophrenia.

The dreams that follow upon death came and subsided, like fibrillar twitchings
Of the nerves unorganizing themselves; and some of the small dreams were delightful and some, slight miseries,

But nothing intense; then consciousness wandered home from the cell
to the molecule, was utterly dissolved and changed.

The complex molecules associated with life decay to simpler, more elemental forms. Because the dialogue that the brain carries on is with itself, this atomic view of death is anti-heroic. Jeffers also refers to dreams after death in the poem "Cawdor."

"Margrave" ends as it began, with the narrator gazing at the stars.

I seem to have stood a long time and watched the stars pass.
They also shall perish I believe.
Here to-day, gone to-morrow, desperate wee galaxies
Scattering themselves and shining their substance away
Like a passionate thought. It is very well ordered.

"Shining their substance away" is literally true because some of a star's mass, m, is converted by means of nuclear reactions into radiant energy, E, according to Einstein's mass-energy equivalence, $E = mc^2$, where c is the speed of light. Since the speed of light is so great, it takes only a small mass consumption to give rise to large amounts of radiant energy. The astronomical ending of the poem makes the reader grateful that the malignant consciousness of the human race can damage only a small part of the vast universe.

Jeffers' reputation as a poet has fluctuated, peaking between 1925 and 1935, and reaching its nadir in 1948 with the publication of *The Double-Axe and Other Poems*. Post-war America was not ready to tolerate a poet whose work transcended Allied loyalties. But a recent reviewer, Peter Brunette, praises the power and authenticity of Jeffers' short lyrics, and says that his literary stock appears on its way up. Some readers—humanists for whom man occupies center stage—allow Jeffers' alleged inhumanism to color their reaction to his poetry. For instance, the critic H.H. Waggoner complains that Margrave is not an interesting character, because no character who in himself illustrates the belief that values are illusory or purely subjective and consciousness unimportant can be interesting.

As an antidote to human megalomania, Jeffers points to the splendors of nature, and reminds us of our smallness and ephemerality. Moral values stemming from the beauty of the universe are for him a source of ennobling strength. He has written as follows:

I think one may contribute (ever so slightly) to the beauty of things by making one's own life and environment beautiful, so far as one's power reaches. This includes moral beauty, one of the qualities of humanity, though it seems not to appear elsewhere in the universe. But I would have each person realize that his contribution is not important, its success not really a matter for exultation nor its failure for mourning; the beauty of things is sufficient without him.

(Quoted in Sister Mary James Power's *Poets at Prayer*)

As an example of Jeffers' moral concerns, consider the poem, "Moon and Five Planets," which links the fall of Finland, spring wildflowers, sunset, and a planetary conjunction.

Five planets and a brilliant young moon
Reach like a golden ladder from the saffron-lined sea-rim
High up the dark blue dome of heaven.

From the astronomical data alone, one could identify the day and year, but the poet gives the date as March 10, 1940. Chaucer refers to a similar conjunction in *Troilus and Criseyde:*

The bent moon with her horns pale,
Saturn, and Jove, in Cancer joined were.

Calculations show a conjunction of the moon, Saturn, and Jupiter in the constellation of Cancer in May, 1385, dating the composition of the poem. "Moon and Five Planets" concludes as follows:

. . . Finland today
After all her winter valor and the great war in the snow,
Is beaten down by machines and multitude.
It will be long before the moon and five planets meet again;
And bitter things will have happened; not worse things.

(from *Be Angry at the Sun,* 1941)

The juxtaposition of the remote astronomical event and the fall of Finland makes each seemingly incongruous incident seem more intense by comparison. From the perspective of the infinite universe the fall of Finland is insignificant, but the poet tempers this thought in the last line with a moral and compassionate comment—nothing worse can happen.

George Gamov and colleagues began several decades ago to investigate theoretically the "big bang" theory of the universe—that the entire universe stemmed from one primeval explosion during which all the elements were formed in the first few seconds. In 1965 when scientists at Bell Telephone Laboratories discovered a low-energy background radiation, they interpreted it as a remnant of the initial explosion, giving credence to the "big bang" theory. Scientists now wonder if the explosive expansion of the universe will someday halt due to mutual gravitational attraction of its members and fall in on itself. After admitting that the "big bang" is indescribable, Jeffers attempts a description in "The Great Explosion" (1962):

. . . There is no way to express that explosion; all that exists
Roars into flame, the tortured fragments rush away from each other into all the sky, new universes

Jewel the black breast of night; and far off the outer nebulae like charging spearmen again
Invade emptiness.
No wonder we are so fascinated with fire-works
And our huge bombs; it is a kind of homesickness perhaps for the howling fire-blast that we were born from.

But the whole sum of the energies
That made and contained the giant atom survives. It will gather again and pile up, the power and the glory—
And no doubt it will burst again; diastole and systole: the whole universe beats like a heart.

Jeffers' explanation for man's propensity for blowing up others of his kind is perhaps farfetched, but not much more difficult to believe than the theory that our aggressive instincts are inherited from the apes. He also refers to one of the great principles of physics and chemistry, that of Conservation of Energy or First Law of Thermodynamics. The sum of all the energies in the universe cannot be diminished or increased, but one kind of energy can be transformed into another kind as potential energy is changed into kinetic energy in the falling of an apple. "The power and the glory" is a reference to the Lord's Prayer. He likens the expansion and contraction of the universe to the beating of a heart, a vivid metaphor for a physical theory.

In "Nova" (1937) Jeffers describes a second-magnitude nova which appeared in June, 1936, and was visible to the naked eye for nine or ten nights.

That nova was a moderate star like our good sun; it stored no doubt a little more than it spent
Of heat and energy until the increasing tension came to the trigger-point
Of a new chemistry; then what was already flaming found a new manner of flaming ten-thousand fold
More brightly for a brief time; what was a pin-point fleck on a sensitive plate at the great telescope's
Eyepiece now shouts down the steep night to the naked eye, a nine-day super-star.

Jeffers is precisely right about the "new chemistry." According to present-day physical theory the stages of evolution for a star not much bigger than our sun proceed as follows: 1) with the consumption of a large part of the star's hydrogen fuel, the star expands to a "red giant" (diameter 250 times that of the sun), 2) with further consumption of helium and heavier elements it collapses to a "white dwarf" (diameter 1/100 that of the sun), 3) the "white dwarf" explodes into a nova, and 4) the remnants of the star collapse to a diameter of perhaps 10 kilometers, forming a "neutron star." J. Robert Oppenheimer and collaborators pre-

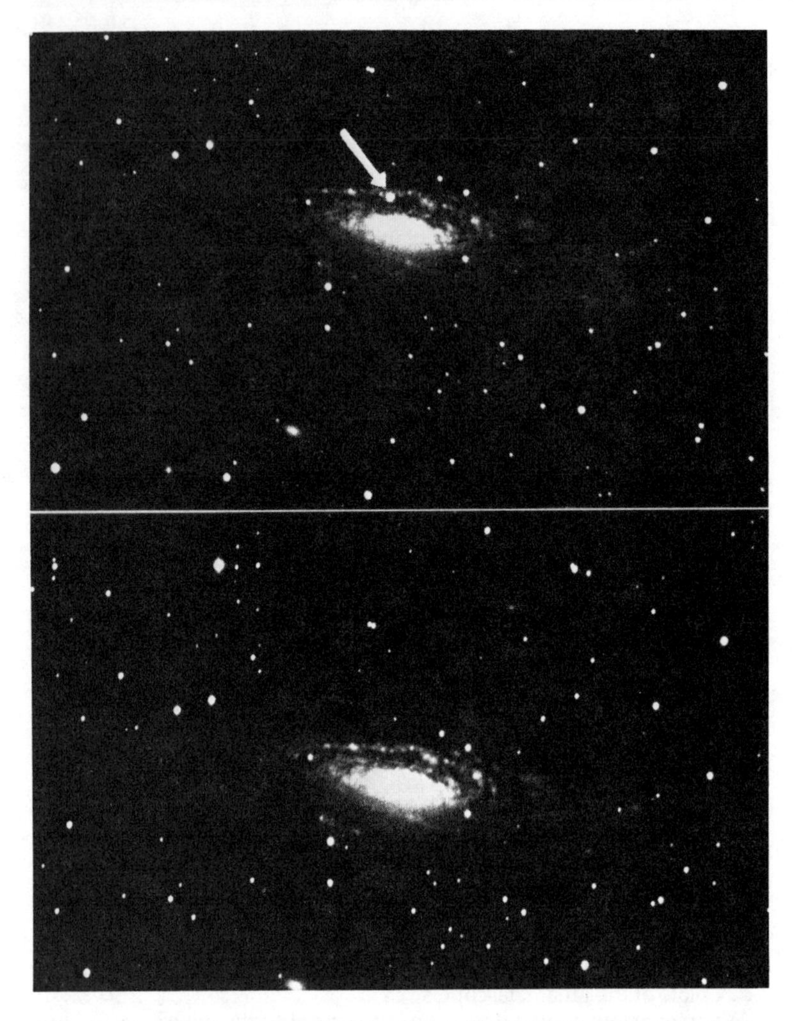

Fig. 22. A supernova (arrow). Two views of NGC 7331, before and during maximum of supernova of 1959.

dicted neutron stars on theoretical grounds in work published in 1938 and 1939, but astronomers did not find experimental evidence of their existence in the form of "pulsars" until 1967. Believed to be rapidly rotating neutron stars, these objects emit regular optical and radio pulses. In the middle section of the poem Jeffers describes what would happen if the sun became a nova.

> . . . The oceans would explode into invisible steam,
> The ships and great whales fall through them like flaming meteors into

the emptied abysm, the six mile
Hollows of the Pacific sea-bed might smoke for a moment. Then the earth would be like the pale proud moon,
Nothing but vitrified sand and rock would be left on earth.

Does this possibility paralyze our actions and render our moral strivings meaningless? Not according to Jeffers; he ends the poem with a beautiful expression of his creed.

We can, by force and self-discipline, by many refusals and a few assertions, in the teeth of fortune assure ourselves
Freedom and integrity in life or integrity in death. And we know that the enormous invulnerable beauty of things
Is the face of God, to live gladly in its presence, and die without grief or fear knowing it survives us.

After "Nova" was published, scientists learned how to use the "new chemistry" to create atomic bombs. Jeffers refers to this nuclear chemistry in the poem "Fire."

There is another nature of fire; not the same fire,
But the fire's father: "Holy, holy, holy,"
Sing the angels of the sun, pouring out power
On the lands and the planets; but it's no holier
Than a fire in a hut, it is another chemistry,
More primitive, more powerful, more universal, power's peak.
. . . . .
It is with this kind of fire
Our people are playing tricks and will blast their enemies.
How brave we are! I would rather think of the little bale-fires
That flickered around the night-gleaming Aegean on the peaks of promontories,
Telegraphing Troy's fall: and that grim Queen had her axe whetted
When the commander of the armies came hulking home.

If the fall of a small city such as Troy had such repercussions, how much greater will be our doom who presume to destroy entire nations. The poem concludes with an image of terror, the black dog.

Fire answered fire,
Blood cried for blood; crime and reprisal, the bomb and the knife, echo forever
To no atonement:
Until annihilation comes leaping like a black dog and licks the dish clean: that is atonement. (from *Hungerfield,* 1951)

Kenneth Rexroth

What are we to make of Kenneth Rexroth, the largely self-educated polymath, political radical, and member of the Beat Generation in San Francisco? Perhaps it is too early for reliable critical judgements to be made, but I rate him as one of the great

Fig. 23. Lucy's reaction to astronomical allusions in poetry. © 1969 United Feature Syndicate, Inc.

poets of our day. His shorter poems offer abundant evidence of his keen interest in astronomy, mathematics, geology and biology. The discussion here will emphasize the first two subjects.

Rexroth's first book of poetry, *In What Hour* (1940), contains a wealth of accurate astronomical and physical references—not merely simple, direct nature poetry, but often coupled with a philosophic quest or response to a social crisis as in "Hiking on the Coast Range." The meaning of this poem, written on the anniversary of the deaths of two men killed in the San Francisco general strike, hinges on a bit of glass apparatus designed expressly for physics demonstrations—Prince Rupert's drops, small teardrop-shaped pieces of glass supercooled at manufacture by quenching in cold water. The resulting internal stress frozen into a drop by this technique of manufacture is invisible until a tip is broken, causing disintegration into fine glass dust. In the Rexroth poem these glass drops are metaphors for drops of blood. The inner constituents of blood—Rexroth mentions patellae, leucocytes, fibrin, and serum—are as hidden from the eye as the motivations of the two strikers until all becomes clear after their blood splatters the pavement.

Implacable and remote heavenly bodies form a poignant contrast to the "nightmare, the dead flesh" of the Spanish Civil War in the poem, "Requiem for the Spanish Dead." An airplane moves in the direction of Hyades (a group of five stars in the constellation Taurus), disappearing beneath the feet of Orion. In winter these constellations would be best situated for early evening viewing (around 9 P.M.) during the month of January. So the airplane, if flying overhead, must be going south. The poem closes with mention of the three brightest stars in Orion (Rigel, Bellatrix and Betelgeuse) and "the great nebula glimmering in his loins" (see Fig. 18). Orion and Taurus also figure prominently in "A Letter to Yvor Winters" in which the Great Nebula is compared to the "clouds of unknowing" of mysticism.

A mention of Deneb, the brightest star in the constellation Cygnus, in the poem, "Climbing Milestone Mountain, August 22, 1947," serves to contrast the inexorable workings of the heavenly spheres with the horrors of the trial and execution of Sacco and Vanzetti. One can actually pinpoint the time of the poem from the line, "We sat up late while Deneb moved over the Zenith." A star chart shows that Deneb reaches its zenith on that date between 10 and 11 P.M.

The poems discussed so far reveal the outrage Rexroth felt when confronted by tragic human destruction. Libertarian and humanistic struggles are conspicuously absent from his poem, "Toward an Organic Philosophy." In this poem the constellations are animated: "Scorpio rises late with Mars caught in his claw," "Orion walks waist deep in the fog," and "the Great Bear kneels on the mountain." The poem closes with an imaginary quotation from John Tyndall, the Irish physicist and alpinist, to the effect that the "concerns of this little place" are determined by the obliquity of the earth's axis. He mentions a "chain of dependence," a reference to the Great Chain of Being linking planet, man and marmot. Perhaps Rexroth was camping in Sequoia National Park near Mt. Tyndall, a 14,025 foot peak named for the scientist.

Many astronomical images appear in the poem, "Easy Lessons in Geophagy." We read of Leonids (a meteor shower), Canopus (a star in the constellation Carina), comet traps, and a meteorite. Rexroth ends the poem with a clever use of the red shift as a metaphor for aging.

> Light is reddened by age, it loses energy as it gets older traveling
> through space.

The astronomical references continue in a companion piece, "A Lesson in Geography." The poet mentions the Great Bear, Algol (an eclipsing variable star—actually a double star with two members revolving around a common center), the constellation Boötes, Sirius (the brightest star in the heavens, located in Canis Major), and Betelgeuse. He explicitly refers to the variable nature of Algol in "Memorandum" from the book, *The Art of Worldly Wisdom*, by calling it "pulsing Algol."

The final poem of *In What Hour* contains numerous scientific references, including mention of Altair (a double variable star in the constellation Aquila), Milky Way, sun, stars, moon, meteors, eclipse, Polaris (the north star), and the constellation Gemini.

Fascinated by mathematics (after all, both poetry and mathematics employ powerful symbols in constructs), Rexroth often introduces a book of poems with a quote from a mathematician. *In What Hour* begins with a sentence from Albert North Whitehead. The lead verse in *Natural Numbers* (1964) is a "found" poem, "A Lemma by Constance Reid," from her book, *From Zero to Infinity: What Makes Numbers Interesting* (Thomas Y. Crowell Co., New York). He prefaces *Gödel's Proof* (1965) with a quotation from the mathematician Kurt Gödel ("A self-contained system is a contradiction of terms. QED"), probably indicating an abandonment of any attempt by the poet to develop an all-embracing philosophy of life.

Mathematical terms abound in "Theory of Numbers" from *The Phoenix and the Tortoise* (1944). Examples of mathematical nomenclature used metaphorically in the poem are the following: space, infinite, points, lines and surfaces, unfolding, limit, differential geometry, existence of integrals, equation, and hexagons. Especially striking is his use of the word "lens."

> This is the lens of intention,
> Focusing liability
> From world to person, from passion
> To action;

Fig. 24. Photograph of Kenneth Rexroth by Margo Moore. New Directions Publishing Corp.

Here the process by which abstract thought converges to personal action is symbolically paralleled with the action of a lens in focusing light. From the same book, the poem *Pliny–IX,XXXVI–Lampridius–XXIX* looks back longingly to an age in which philosophy, mathematics and astronomy were all part of a common culture.

Before breakfast . . . philosophic
Discourses in the bath, flute players
For lunch, along with mathematics,
Roast peacocks for dinner, and after,
Mixed maenads, or else astronomy
Depending on the mood and weather.

"OTTFFSSENTE," a poem from *Gödel's Proof* (1965) in the so-called "cubist" or fragmented style, contains numerous references to mathematical and other kinds of series, including the famous doggerel, "ABCD goldfish." Even the title is a series, albeit a simple one—merely the first letters of the numbers one through eleven. The giveaway here is that the first word of the poem is "twelve." Implying that a correct and complete world view is impossible, Rexroth writes the following mathematical lines:

to the mathematically mature it is well known that there
is no such thing as the correct missing number in any
specified sequence. It is possible to insert any number whatever
and find a formula which will justify the sequence term for term.

Other poems with references to mathematics include "Death, Judgement, Heaven, Hell," "Inversely, As the Square of their Distances Apart," and "Time is an Exclusion Series Said McTaggart."

Rexroth's book and poem "The Signature of All Things" (1949) take their name from the book by the seventeenth-century mystic Jacob Boehme.

The saint saw the world as streaming
In the electrolysis of love.

In Rexroth's brilliant metaphor the streams of bubbles of oxygen and hydrogen from the electrolysis of water stand for the light of divine love streaming through the universe. In one case the source is electrical energy, and in the other, the divine creative spark.

"The Lights in the Sky are Stars," written for his four-year-old daughter Mary, is Rexroth at his most charming. It uses astronomical references to set the time and to emphasize by contrast the hurly-burly of earth. He mentions Halley's comet, the star cluster in the constellation Hercules, Deneb, the Beehive in the Crab (a star cluster in the constellation Cancer), Orion, moon, sun, and various planets. He combines astronomy with biology in the following metaphor:

I pick up the glass
And watch the Great Nebula
Of Andromeda swim like
A phosphorescent amoeba
Slowly around the Pole.

In the section of the poem called "Protoplasm of Light," after an extended description of his experience of a solar eclipse, Rexroth shifts the reader's thoughts to a future when scientists will have fully understood all the phenomena associated with an eclipse.

> The mysterious
> Cone of light leans up from the
> Horizon into the pale sky.
> I say, 'Nobody knows what
> It is or even where it is.'

What Rexroth saw was most likely the zodiacal light, an oblique pyramid of feeble light strongest in the vicinity of the sun and rising from the horizon along the line of the great circle running through the constellations of the zodiac. One can see the zodiacal light in the spring beginning about two hours after sunset in the west, in the autumn before sunrise in the east, and during a solar eclipse. It may be caused by a disc of cosmic dust, within the orbit of the earth, which scatters the light of the sun. Or perhaps it may represent a more complex interaction of the solar wind with dust particles.

Similarities exist between the work of the American poet Robinson Jeffers and Rexroth. Both look up at the stars from the western edge of the continent, but to go from the poetry of Jeffers to that of Rexroth is like coming in from the cold. Jeffers bitterly refers to the galaxies fleeing the "contagion of consciousness," while Rexroth compares Orion to

> . . . some immense theorem
> Which, if once solved, would forever
> Solve the mystery and pain
> Under the bells and spangles. ("The Lights in the Sky Are Stars")

Not only does Rexroth have faith in human reason, but some of his poems indicate a faith in God—not the traditional god of his parents, but the more abstract God of the Christian mystics, Martin Buber, and Buddhism. When Rexroth approaches the transcendent, he cannot resist astronomical associations, as in the poem "Phaedo."

> After Midnight Mass
> In the first black subzero hour of Christmas
> I take a twig and white piece of paper
> And show you the fragile shadow of Sirius
> The Dog Star guarding the Manger
> Sleeping at the foot of the Cross

Jeffers' misanthropic pessimism leads him to apocalyptic visions of nuclear annihilation in the poem "Fire." In contrast, Rexroth suggests in "Yin and Yang" that we may expect cycles of death and rebirth in human society as well as in nature.

While the above citations are not comprehensive, I have given enough samples of Rexroth's usage of naked-eye astronomy and elementary mathematics and physics to demonstrate that these scientific references form an integral part of his poetic output. Incredibly, in a book of literary criticism devoted to Rexroth, Morgan Gibson devotes only one paragraph to science. I think that I have demonstrated that Rexroth on science deserves to be taken more seriously than that.

John Ciardi and Others

John Ciardi's pithy comment on the space age is typical of his work: facile, but with hints of deeper meaning.

First a monkey, then a man.
Just the way the world began.
("Dawn of the Space Age" from *In Fact,* 1962)

It suggests that the human race is in charge of its own evolution, and also makes one wonder if there will be a repeat of Adam and Eve and the Fall of Man. Will we spread our garbage to the stars? Ciardi's light touch strikes just the right note for a disillusioned and cynical twentieth-century audience impatient with those claiming to have all the answers.

Ciardi leads the reader through an imaginative re-creation of the Ptolemaic universe in "My Father's Watch."

One night I dreamed I was locked in my Father's watch
With Ptolemy and twenty-one ruby stars
Mounted on spheres and the Primum Mobile
Coiled and gleaming to the end of space.

He marvels at the "infinite order," but is curious to find out "What makes it shine so bright." The system self-destructs when he "touched the mainspring."

And woke on a numbered dial where two black swords
Spun under a crystal dome. There, looking up
In one flash as the two swords closed and came,
I saw my Father's face frown through the glass.
(from *As If,* 1955)

"Father" of course refers both to the poet's father and to God. The poem as a whole recalls the Deists' arguments that the universe was so like a clockwork mechanism that it must have had a Maker. It suggests that scientists have damaged this beautiful conception by their insatiable curiosity. The poem, "Some Sort of Game," should be read as a companion piece. Here Ptolemy is likened to a toy-maker.

Toy-maker Ptolemy
made up a universe.

Nine crystal yo-yos he
spun on one string. It was
something to see it go,
half sad to see it pass.

Just as the adult looks with nostalgia on the toys of childhood, so the poet views the passing of the Ptolemaic conception. He also suggests that science has altered the conceptions of religion since telescopes cannot be used to view angels.

Somewhere among
such angels as men hope
are there, but do not know,
a dark began to grope.
So toys and angels go. (from *This Strangest Everything,* 1966)

Playing with toys can be dangerous, as Giordano Bruno found out. That philosopher lectured on the Copernican theory, and proclaimed that the universe was infinite and contained stars like the sun with orbiting planets. In 1592 he was charged with heresy, and in 1600 was burned at the stake by the Inquisition. He is memorialized in Ciardi's "For Giordano Bruno's Monument in Campo dei Fiori" (from *39 Poems,* 1959). Ciardi's poem is notable for its verbal inventiveness ("manskin," "manface," "grave-spill") and for its final triumphant image:

Bursting from carrion like a gull surprised—
The sprung word's wingspread taking all the air.

Ciardi is not sentimental about the corpse; more important is the imperishable truth of Bruno's teachings.

In "Galileo and the laws" Ciardi plays upon double meanings of words. Galileo is at first blind to the deeper recesses of human psychology, later he becomes physically blind. Ciardi capitalizes the word "law" when it refers to the inner workings of the universe, and does not when it refers to human rules of conduct.

Galileo thought he saw
the spinning center of the Law
radiating pure equations.
He did, too. But the coruscations
of that center left him blind
to the dark rims of man's mind.

. . . . .

In his old age he knelt and lied
the law away. Before he died,
Milton called, and wept to find
Galileo had gone blind. (from *This Strangest Everything,* 1966)

Although he is better known for his novels, another contemporary American writer, John Updike, has published some largely superficial, mostly humorous verse incorporating scientific concepts. His longish poem, "Midpoint," has a middle section, "The Dance of the Solids," filled with the language of solid state physics from the September 1967 issue of *Scientific American*. "Cosmic Gall," a light verse about neutrinos, ends with the following thought:

At night, they enter at Nepal
And pierce the lover and his lass
From underneath the bed—you call
It wonderful; I call it crass. (from *Telephone Poles and Other Poems*)

In a more serious vein than Updike are the poems of the distinguished authors, Archibald Macleish and Marianne Moore. Readers may wish to pursue Macleish's concept of space-time in the poems, "Signature for Tempo," "Einstein," "Reply to Mr. Wordsworth," and "Seeing." Moore's thoughts on time are expressed in her poem, "Four Quartz Crystal Clocks."

Conclusion

During the century considered here poetic themes accurately reflected scientific developments. Whitman enthusiastically embraced Darwinian evolution; to him it was the scientific foundation for his profound democracy. Frost limited himself to such scientific metaphors as might occur to a "literate farmer." The post-war poets had to face the possibility of nuclear annihilation. Ciardi handles the subject gingerly, but Jeffers tackles it head-on. Jeffers' images may not be likeable, but they are unforgettable ("annihilation . . . leaping like a black dog"). Man's actions pale into insignificance when viewed against the background of the immensity of the universe. These American poets, with the possible exception of Frost, have been "harbingers," "mirrors," and "barometers" to the scientific tempers of their age. The challenge to contemporary poets is to digest an increasingly abstruse modern science and to make it an organic part of their poetic conceptions.

Bibliography

Beaver, Joseph, *Walt Whitman: Poet of Science* (Octagon Books, New York, 1974).

Berman, Louis, "The Wayward Heavens in Literture," J. Col. Sci. Teaching **5,** 82, 1975.

Gibson, Morgan, *Kenneth Rexroth* (Twayne Publishers, Inc., New York, 1972).

Hiers, John T., "Robert Frost's Quarrel with Science and Technology," Georgia Rev. **25,** No. 2, 182, 1971.

Pearson, Norman H., "The American Poet in Relation to Science," Amer. Quart. **1,** No. 2, 116, 1949.

Scott, Robert Ian, *Robinson Jeffers' Poetic Use of Post-Copernican Science,* Ph.D. dissertation, State Univ. of N.Y. at Buffalo, 1964.
Waggoner, H.H., *The Heel of Elohim: Science and Values in American Poetry* (Univ. of Oklahoma Press, Norman, 1950).

Chapter 6

Acoustics of Music

MUSICAL ACOUSTICS comprises the fundamentals of production of musical sound, musical instrument design, and architectural acoustics of concert halls. Each of these areas of musical acoustics will be treated in the following sections, using the piano as an illustration of a typical sound-producing device. Of all the instruments of the orchestra I have chosen to spotlight the piano here not only because it is readily available in many homes for inspection and experimentation, but also because many general principles applicable to other musical instruments will emerge from its study. Such a study is not essential to the enjoyment of music. Obviously, millions of people enjoy piano recitals with only the vaguest notion of what musical acoustics is all about, but contemplation of the centuries of human endeavor resulting in an understanding of the intricacies of vibrating string motion, piano construction, musical scales, and concert hall acoustics adds another dimension (for me a very important one) to the pleasure of musical listening.

A Vibrating String

Stringed instruments are among the oldest musical instruments depicted in tomb drawings and reliefs. For example, lyres were used in Sumerian music, as shown in reliefs from a royal tomb at Ur (2700 B.C.). Pythagoras initiated a mathematical approach to the subject of the vibrating sring some 2500 years ago by experimenting with a single stretched horizontal string attached at one end to a peg and weighted at the other end by a stone hung vertically from the end of the instrument. By using different sizes of stones he could adjust the *tension* or force (measured in pounds) on the string. He noted that the pitch of sound rose when he shortened the string while maintaining the same tension. By dividing the string into two vibrating sections with a bridge he found pleasing combinations of sounds when the two lengths were in the ratio of small whole numbers.

By Galileo's time the concept of pitch had been associated with *frequency*, a term to be defined below. A *sine wave* represents the simplest type of vibration (see Fig. 25). Musical instruments rarely emit a single pure sine wave, but an electronic audio oscillator does. Connect a loudspeaker to an electronic audio oscillator; the resulting sound has a machine-like and inhuman quality. A graph of the sine wave shows that it repeats itself or cycles at an interval of time called the period, T. Frequency, the number of cycles of vibration per second, is therefore the inverse of the period. The modern name for cycle per second is "hertz" (abbreviated Hz), after Heinrich Hertz, the discoverer of sinusoidal radio waves. High-pitched sounds have large frequencies of vibration and small periods, while the opposite is true of low-pitched sounds.

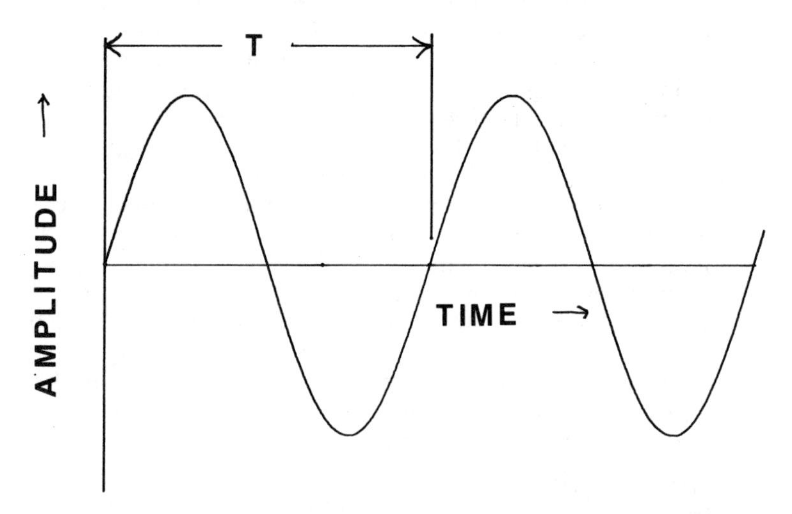

Fig. 25. An amplitude-time graph of a sine wave. The time for one complete cycle is the period, *T.*

Galileo, his curiosity aroused while playing the lute, observed that the frequency of a vibrating elastic string fixed at both ends is inversely proportional to its length, ℓ, and proportional to the square root of the tension, T, divided by the mass per unit length, ρ,

$$f = \frac{1}{2\ell}\sqrt{\frac{T}{\rho}}.$$

Later, eighteenth-century mathematician Brook Taylor mathematically derived this equation from first principles. The problems of describing the displacement of a string as a function of time starting from an arbitrary shape engaged some of the most brilliant mathematical minds of the eighteenth century: John Bernoulli and his son Daniel, Jean Le Rond d'Alembert, Leonard Euler, and Joseph-Louis Lagrange. Their efforts to describe all possible motions of a string led to the creation of a new branch of mathematics, partial differential equations. They proved that a string vibrates at its fundamental frequency simultaneously with a number of higher frequencies called *overtones*. Overtones with frequencies 2, 3, 4, . . . times the fundamental frequency are called *harmonics*. The sound of an ideal vibrating string is a mixture of the fundamental frequency plus its harmonics.

The ideal string, an abstraction of pure mathematics, has no stiffness and perfect uniformity of diameter and density. When disturbed from its rest position, the ideal string returns to its original profile by tension applied to the ends of the string with no contribution from the stiffness of the string. Since all real strings are stiff to some degree, real vibrating strings do not emit a pure harmonic sequence of overtones. Tension dominates over stiffness in a case approximating ideal. When stiffness predominates (as in a slack string or a bar clamped at both ends), a vibrating string emits a very non-musical sound.

If one asks orchestra players to sound the same note on different instruments in turn, anyone who has paid close attention to Prokofiev's *Peter and the Wolf* or Britten's *Young Person's Guide to the Orchestra* can readily identify the various instruments. Each instrument emits the same fundamental frequency, but differing combinations of harmonics. The harmonic recipe gives each instrument its unique coloration or timbre. But attempts to duplicate electronically the sound of a piano by combining the fundamental frequency and its recipe of overtones with a piano-like attack and decay of sound have failed. One must add certain mechanical noises arising from the hammers, pedals, and dampers to achieve a convincing piano tone.

The mathematicians came very close to proving that the general solution to their problem was the sum of an infinite series of trigonometric functions. Joseph Fourier discovered the elusive proof in the nineteenth century. He was not studying vibrating strings at the time, but instead trying to solve a problem dealing with flow of heat. With this culminating achievement, the theory of ideal vibrating strings was essentially complete.

The work on the vibrating string problem set the stage for future advances in physics in areas that the mathematicians of the time could not have anticipated. Because of the earlier study of a vibrating string, the mathematical apparatus of differential equations was readily available when Erwin Schrödinger developed his revolutionary theory of wave mechanics in 1926. For example, one need not know any mathematics to perceive the formal similarity between the equation for a vibrating string,

$$\frac{d^2\psi}{dx^2} + \left(\frac{\rho\omega^2}{T_0}\right)\psi = 0 ,$$

and the time-independent Schrödinger equation,

$$\frac{d^2\psi}{dx^2} + \left(\frac{2mE}{h^2}\right)\psi = 0 ,$$

even though the constants (symbols in parentheses) multiplying the second term are different. The first equation describes the amplitude of vibrations of a string, while the second could describe the wave function (its amplitude squared is proportional to the probability of finding a particle in a particular place) of an electron in a closed box. When one solves the equation, one finds the variable psi (ψ) as a function of x. One graph (Fig. 26) could represent the solutions of both equations with suitable redefinitions of the variable displayed on the ordinate. This graph represents either a string of length *a* vibrating at its fundamental frequency (ψ_1), and second (ψ_2) and third (ψ_3) harmonics, or the allowed wave functions of an electron confined to a one-dimensional box of length *a*. the points along the horizontal axis where the wave amplitude is always zero are *nodes,* and those points which have maximum or minimum amplitudes are *antinodes.*

It has not always been possible for mathematicians to discover a theoretical explanation for every aspect of the complicated behavior of vibrating systems. Modern investigators with audio frequency analyzers and digital computers have been exploring finer and finer details of the sound of actual piano strings. They have found that the amount of departure from a harmonic series increases the higher in frequency an overtone becomes. In addition the overtones of a given note vary greatly in loudness, and each overtone decays in time at a different rate. At present these phenomena remain a challenge to theoreticians.

Early Stringed Instruments with Keyboards

The mathematicians' struggles to understand in detail vibrating string motion had little effect on the evolution of the piano, which proceeded by trial and error. Descriptions of a keyboard instrument known as the *clavichord* (from the Latin words for *key* and *chord*) date from the first half of the fifteenth century. When one depresses a key of this instrument, a wedge-shaped metal striker or *tangent* presses upward against the string, dividing it

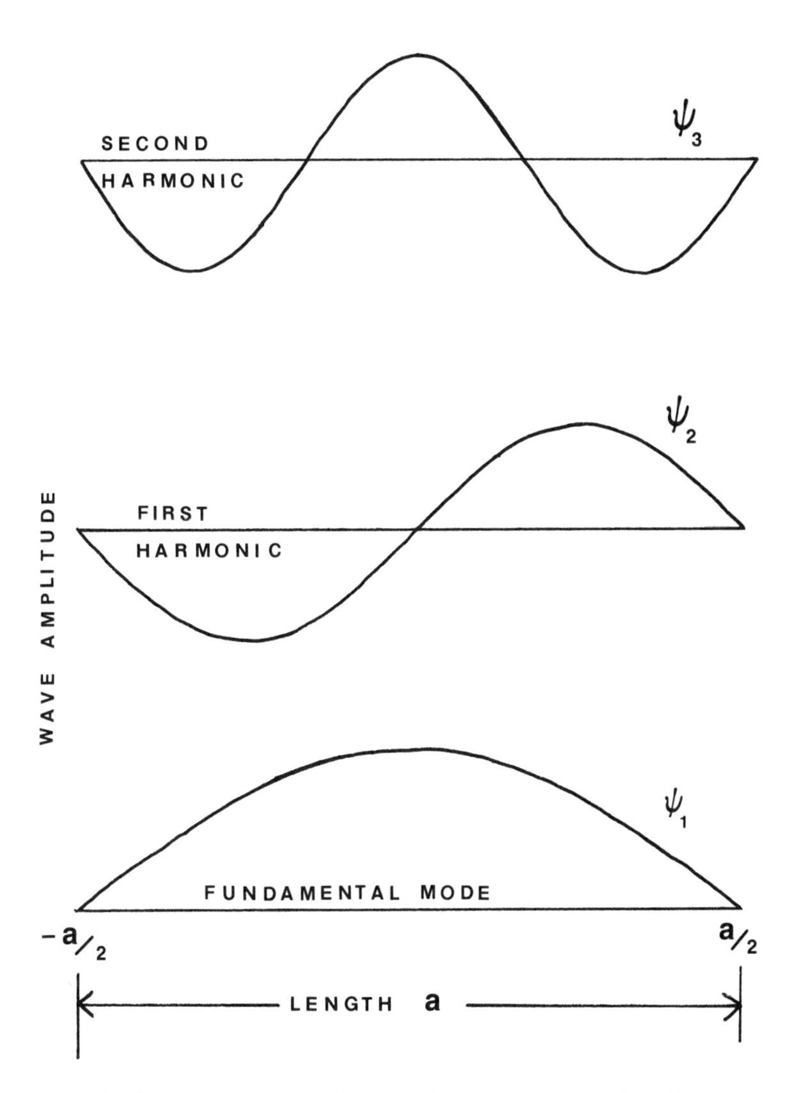

Fig. 26. Sine waves representing, from bottom to top, a string of length a *vibrating at its fundamental frequency, and second and third harmonics. Alternatively, the diagram could represent the allowed wave functions of an electron confined to a one-dimensional box of length* a.

into two parts (Fig. 27a). One part of the string vibrates with the tangent acting as one of the bridges, while strips of felt damp the other segment. When the player releases a key, the felt strips damp the entire string. Since the tangent remains in contact with the sounding string, the player has a great deal of control over the

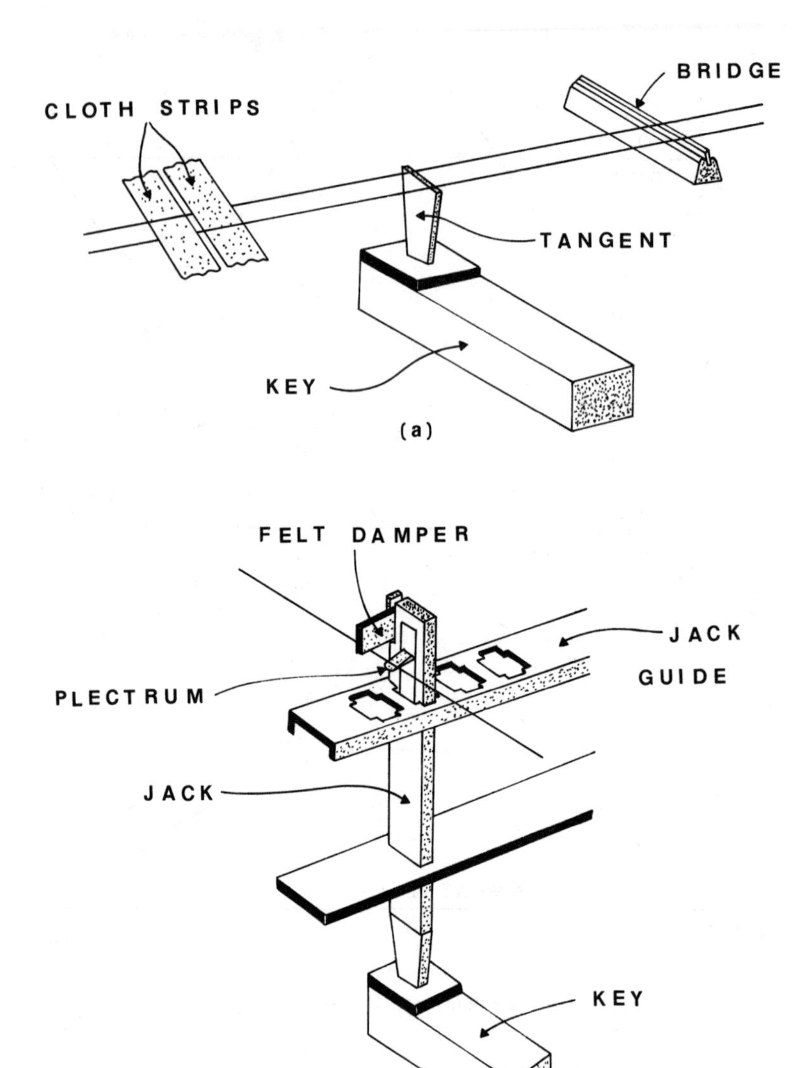

Fig. 27. The action of (a) a clavichord, and (b) a harpsichord.

sound quality within a rather limited dynamic range. If one varies finger pressure while depressing a key, the effect is a kind of vibrato. A firm touch can vary the pitch of a note because the pressure of the tangent against the string changes its tension. The clavichord at its loudest is hardly as loud as a piano at its softest, because the manner of striking a string in the clavichord limits its loudness. Striking a string at one end (a node for the fundamental frequency and all the even harmonics) is an extremely inefficient way to impart vibrational energy to it. Since its tone is so soft, the clavichord tends to be drowned out if used in ensemble music.

An increase in loudness was possible with the *harpsichord* (from the Latin words for *harp* and *chord*), an instrument popular from the early sixteenth to late eighteenth centuries. In the harpsichord the strings are plucked rather than struck as in the clavichord. A vertical wooden *jack* rests on the key (Fig. 27b). A wooden jack holds the plucker or *plectrum* (a quill or bit of leather) on a pivoted support, and a felt damper also fitted to the jack in a narrow slot rests on the string when it is not sounding. When one depresses a key, the jack rises and the plectrum forces its way past the string, plucking it. Neither the jack nor the plectrum is in contact with the string while it is sounding. Upon release of the key the plectrum pivots backward so that it can brush past the string without plucking it. The felt damper then silences the string.

The plucking mechanism has the disadvantage of allowing only minor variations in loudness and tone quality to be achieved by the player. To overcome this limitation, harpsichord builders in the middle of the sixteenth century began to produce instruments having two jacks and two strings for each key. A jack guide shifted an entire row of jacks so that a player could adjust the instrument to pluck one or two strings for each key. Such manual adjustments were necessarily slow. During the seventeenth century builders incorporated two keyboards into some of their instruments, one keyboard operating a single row of jacks, and the other two rows of jacks. Thus it was possible to play loudly with one hand and softly with the other.

The Italian harpsichord maker Bartolomeo Christofori, working for the Medici family of Florence around 1700, attempted to make an instrument combining the expressiveness of the clavichord with the loudness of the harpsichord. He called his invention *gravicembalo col piano e forte* (literally, "harpsichord with soft and loud"), hence the name *pianoforte,* or more simply, piano. In Christofori's mechanism, felt-covered hammers hit the strings, rebounding quickly lest they damp the note they were intended to sound. The *action* (keys, hammers, and mechanism for setting the hammers in motion and restraining their bounce) devised by Christofori differs only in detail from a modern piano action illustrated in Fig. 29. Although builders continued to make harpsichords and pianos side by side, the success of the new instrument was so great that by 1800 pianos dominated production.

One must hear the sounds of the early keyboard instruments to appreciate their unique virtues. I can recommend a selection of records, by no means comprehensive, which presents music appropriate for each instrument. A two-record set, *Historical Keyboard Instruments at the Victoria and Albert Museum, London* (Musica Rara MUS70-1) features virginals (a form of harpsichord in which the strings run from left to right, parallel to the keyboard), harpsichords, and pianos from the Victoria and Albert's Music Gallery. Thurston Dart performs an even larger selection of pieces on a five-record set, *Early English Keyboard Music* (L'Oiseau-Lyre OLS114-8). The instruments used include organ, harpsichord, and clavichord. Wanda Landowska performed on a Pleyel harpsichord making numerous recordings. This instrument, produced in Paris, embodied piano construction techniques such as a cast iron frame and one-quarter inch laminated soundboard. It is really a plucked piano, and does not sound at all like the earlier harpsichords. More recently, Igor Kipnis has recorded several discs of clavichord and harpsichord music. Among them are *The Art of Igor Kipnis* (Columbia M3X32325) and *Austrian Music for Harpsichord and Clavichord* (Odyssey Y30289). The pianist Jörg Demus uses a wooden-frame piano from his own collection (a Rausch piano from about 1835) in certain of his Schubert recitals, for example, *The Music of Franz Schubert* (BASF-Harmonia Mundi KHB-21442). I would not cavil at the date of the piano (seven years after Schubert's death); essentially Demus plays Schubert's music on a piano of Schubert's time.

The Design of a Piano

If one takes Galileo's equation (first section of this chapter) as the sole basis for piano design, some problems will immediately be apparent. The frequency range spanned by the keyboard is from about 27 to 3520 hertz. If one starts with the assumption that all strings should have the same diameter (related to mass per unit length) and tension as the ones for the highest notes, and if one notices that the strings for these notes are about two inches long, then a simple use of the equation reveals that the strings for the lowest notes must be nearly 22 feet long! A second attempt at the design of the piano might begin by noting that one could also achieve low frequencies by lowering the tension on a string. Therefore, one may arrange the strings for the treble as before, with the bass strings shortened and loosened. Unfortunately, this reduces the restoring forces necessary for efficient vibration and spoils the sound. In fact, the strings are all under more or less constant tension. The only remaining variable in the equation is the mass per unit length, ρ, which is proportional to the diameter of the string. If one keeps constant tensions on all strings and adjusts the lengths to fit into an ordinary piano, another simple calculation shows that the diameter of the string for the lowest note must be as thick as a pencil! A rod of this diameter, length, and tension would not sound at all like a piano string; it would have a higher tone than predicted from Galileo's equation because of its stiffness, and the departure of the overtones from a pure harmonic structure would be very noticeable. The way in which one solves the problem in practice is to wrap the strings for the notes lower than the octave below middle C with a coil of copper or alloy. This increases the mass per unit length without unduly increasing stiffness. Thus, makers of pianos have had to depart from the ideal case in order to build a piano which could fit into a living room.

One can demonstrate the principal difference in sound between a concert-grand piano and an upright by sounding the bass notes of each. The bass strings of a concert grand are fully twice as long as those of an upright. For the same diameter, the tension on the concert-grand's bass strings may therefore be four times the tension on the upright's strings to achieve the same frequency. Thus the tension-to-stiffness ratio is greatly enhanced for the concert-grand's bass strings, making them closer to ideal strings. The resultant tones are much more musical than the sounds from an upright's nether regions.

Modern piano makers utilize the seventeenth-century harpsichord builders' technique of using more than one string per note to increase volume. Each note comprises from one to three strings tuned in unison and struck simultaneously by one hammer. This stratagem alone would not be enough to achieve the customary loudness from the piano, because the rather thin diameter of a string cannot set a sufficient volume of air in motion. Piano makers overcome this difficulty by communicating the vibrations of the strings to a large wooden soundboard, the "floor" of the piano, which radiates the sound efficiently. The vibrating portion of a string is between the agraffe and an S-shaped wooden bridge attached to the soundboard. When one strikes a key, the sound must of course initially come from the related string; then, after a short interval known as the attack time, during which the vibration is transferred through the bridge to the soundboard, the sound radiates mainly from the soundboard.

Strings could be stretched to moderate tension on the wooden frames of early pianos. With the introduction of a cast-iron frame in 1855, Henry Steinway effected a considerable increase in brilliance and power. The force on one of the 240 individual strings may be as high as 450 pounds, while the total pull the frame must withstand is in the neighborhood of thirty tons for a concert-grand piano. The frame itself weighs 400 pounds and is cast in one piece (see Fig. 28). A hitch pin anchors one end of a string while a tuning pin anchors the other end. One can adjust the tension on each string by turning its tuning pin.

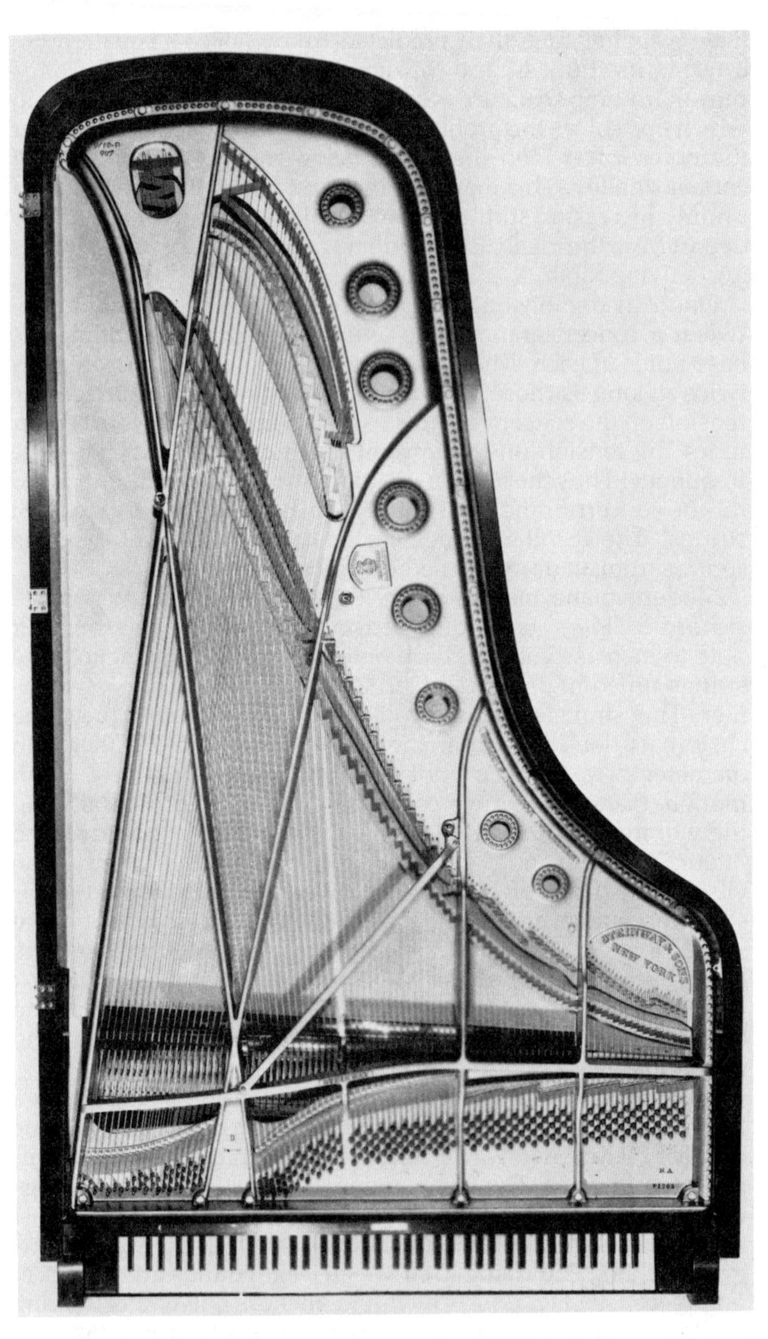

Fig. 28. Vertical view of a concert-grand piano. This is Steinway and Sons' Model D.

A piano case conceals the action, a wonderful example of craftsmanship. This complicated mechanism must perform the following functions: it has to provide free travel of the hammer during that portion of its swing just before it hits the string, it must allow gravity return of the hammer but no rebound which might result in a double note, it must lift a damper (a felt-covered bar which prevents vibration of the string) while the key is depressed, and it must enable a player to play a rapid succession of notes. The hammer head is covered with felt that one may file to make harder or prick with a needle to make softer, producing a mellower tone. The free swing of the hammer before it hits the string has led to a controversy about the effect of the "touch" of a pianist. Scientists have insisted that there could not be any difference between an electrically actuated key and one depressed by a human finger since the key is effectively disconnected from the hammer during the final moments before impact with the string. Indeed, they have submitted oscilloscope pictures of waveforms showing that the two sounds are identical for a given keyboard pressure. Equally insistent have been musicians who heard a difference even if the scientific instruments could not pick it up. A reconciliation of these two views is now possible, because the velocity of the hammer striking a string affects sound quality as well as loudness. If one sounds a single note there can be no difference between the electrical and the human actuator, but if one plays a chord, an artist can vary the force with which individual notes of the chord are struck, and thus vary its timbre.

The sound of a string varies greatly depending upon whether it is plucked with a sharp instrument or hit with various shapes of hammers. In other words, initial conditions are very important in determining the subsequent course of events in vibration. Thomas Young formulated in 1800 a more precise statement of these facts. He showed mathematically that no disturbance applied at a node (recall that this is a continuous zero displacement of the string) of a harmonic can excite that particular harmonic. According to Young's law, a hammer striking a string one seventh of the way along the string could not excite the 7th, 14th, 28th, . . . harmonics. In a piano string, those nodes within the length of contact of the hammer with the string would be missing. The seventh and ninth harmonics are dissonant as will be seen, so the hammers are arranged to strike from one-seventh to one-ninth of the way along the strings.

Young's law partially explains why a soft hammer (pricked felt) gives a different string tone from a hard hammer (filed felt). The soft hammer has a greater width of contact with the string and thus suppresses more overtones than the hard hammer. In addition, another factor is at work. During the instant the hammer touches the string before rebound, the string is vibrating in two parts (clavichord vibrations) which are not necessary musically related to the note to be sounded by the whole string. The felt covering of the hammer damps these undesirable vibrations. A piano tuner can adjust the hardness and breadth of each hammer for the smoothest progression of notes.

Two of the three pedals on a piano control the action of the dampers. Normally the action releases a string damper only during depression of its particular key. The *forte* or sustaining pedal on the right lifts all the dampers so that any of the strings may vibrate sympathetically (resonate) with the one hit by the hammer. This will occur if a particular overtone in the vibration recipe of a string happens to correspond with one of the overtones in the sounding note. The effect is to change the coloration of the tone. One can easily demonstrate sympathetic vibration of strings by depressing the sustaining pedal and singing into the strings of an open piano. When one's voice stops, one can still hear the note emanating from strings vibrating in resonance with the voice. The *sostenuto* or middle pedal lifts the dampers of only those strings which are being struck during the depression of the pedal. These dampers remain lifted until release of the pedal. The pedal on the left does not affect the dampers, but shifts the entire action so that the hammers do not strike all three strings of the same note. This has the effect of decreasing the loudness of a note, or "softening" it.

Musical Scales

The keyboard of a piano contains notes which get progressively higher in frequency from the left end of the keyboard to the right. When one specifies the interval in frequency between one note and the next according to a set formula as in the piano, the arrangement of notes forms a musical scale. Many famous individuals have devised scales: Pythagoras constructed one musical scale and Galileo's father another. The most natural one, in the sense that it evolved anonymously according to the laws of acoustics and the nature of human hearing, is the *diatonic* scale (sometimes called true or "just" intonation).

With a certain amount of guesswork, it is possible to reconstruct some of the steps in the evolution of the diatonic scale. When men and women first began to sing together, they probably sang an octave apart (frequency ratio 1:2), thinking that they were singing in unison, because the natural difference in their voices is about an octave, and because the fundamental vibration and the second harmonic have several nodes in common (Fig. 30). The fifth (frequency ratio 3:2) occurs quite early in written music, attesting to its fundamental nature. Students with no

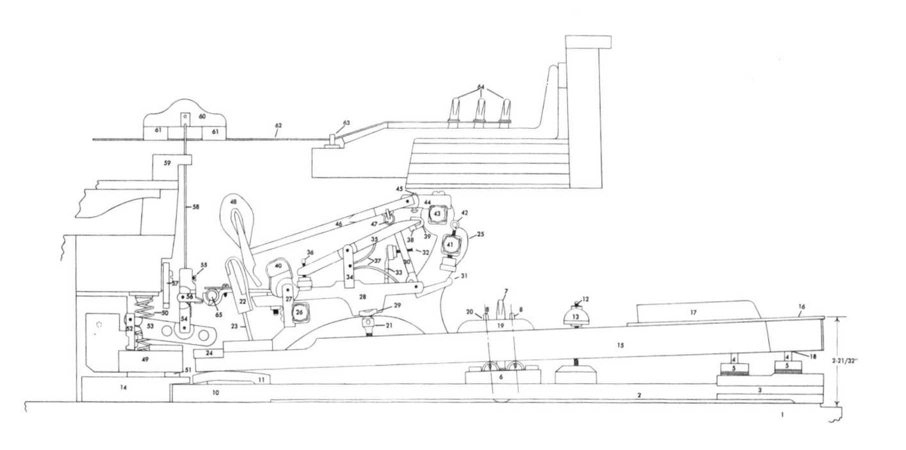

Fig. 29. Diagram of a grand piano action. The names of the numbered parts are as follows:

1 Keybed
2 Keyframe
3 Front Rail
4 Front Rail Pin
5 Front Rail Pin Punching
6 Balance Rail
7 Balance Rail Stud
8 Balance Rail Pin
9 Balance Rail Bearing
10 Back Rail
11 Back Rail Cloth
12 Key Stop Rail Prop
13 Key Stop Rail
14 Dag
15 Key
16 Key Covering
17 Sharp
18 Front Pin Bushing
19 Key Button
20 Balance Pin Bushing
21 Capstan Screw
22 Backcheck
23 Backcheck Wire
24 Underlever Key Cushion
25 Action Hanger
26 Support Rail
27 Support Flange
28 Support
29 Support Cushion
30 Fly
31 Tender
32 Fly Regulating Screw
33 Spoon
34 Support Top Flange
35 Balancier
36 Balancier Regulating Screw
37 Repetition Spring
38 Repetition Felt
39 Balancier Covering
40 Hammer Rest
41 Regulating Rail
42 Letoff Regulating Screw
43 Hammer Rail
44 Hammershank Flange
45 Drop Screw
46 Hammershank
47 Knuckle
48 Hammer
49 Underlever Frame
50 Underlever Frame Spring
51 Underlever Frame Cushion
52 Underlever Flange
53 Underlever
54 Underlever Top Flange
55 Damper Wire Screw
56 Tab
57 Damper Stop Rail
58 Damper Wire
59 Damper Guide Rail
60 Damper Head
61 Damper Felts
62 String
63 Agraffe
64 Tuning Pin
65 Sostenuto Rod

musical background often will tune a sonometer (a single string on a sounding board) and a tuning fork a fifth apart, thinking they are in unison. To Galileo the fifth was even more pleasing to the ear than the octave, because "the sweetness is tempered by a sprinkling of sharpness, giving the impression of being simultaneously sweetly kissed and bitten." So both the octave and the fifth sound relatively pleasant, that is, they form *consonances*. But why should this be so?

To understand the nature of consonance and dissonance one must understand the phenomenon of *beats*. Consider two notes sounded at the same time; the ear and brain perceive their average tone as well as a difference or beat frequency. To hear beats, connect loudspeakers to two audio oscillators and set them a few hertz apart, say 440 and 444 Hz. A beat frequency of 4 Hz will be heard as well as the average frequency of 442 Hz. If a piano is available, a more convenient way to hear beats is to strike a note while humming just off the same pitch. Baron von Helmholtz conducted a series of investigations into the nature of consonance and dissonance (1862). He suggested that the relative unpleasantness of two notes sounded together is due to the presence of beats between the two notes or their harmonics: the greater the number of different beats the greater the dissonance. If the original tones and their beats are not related by small whole numbers, the quality of sound has a roughness unpleasant to most people. The following tables list the fundamental frequency on the left, followed in the same line by its harmonics.

Table 5.

Two Notes an Octave Apart

Frequency of first note and its harmonics	f	2f	3f	4f	5f	6f	7f	...
Frequency of second note and its harmonics		2f		4f		6f		...

Note that in Table 5 all the listed frequencies have in common the fundamental frequency, f. Therefore one could simplify the table by dividing all the expressions by the fundamental frequency. Subsequent tables in this chapter will use the resultant frequency ratios. The whole-number multiples of the fundamental frequency are the harmonic frequencies. Note that all of the harmonics of the higher note are common to both notes. Thus the octave is the most perfect of all concords. Next consider the interval of a fifth.

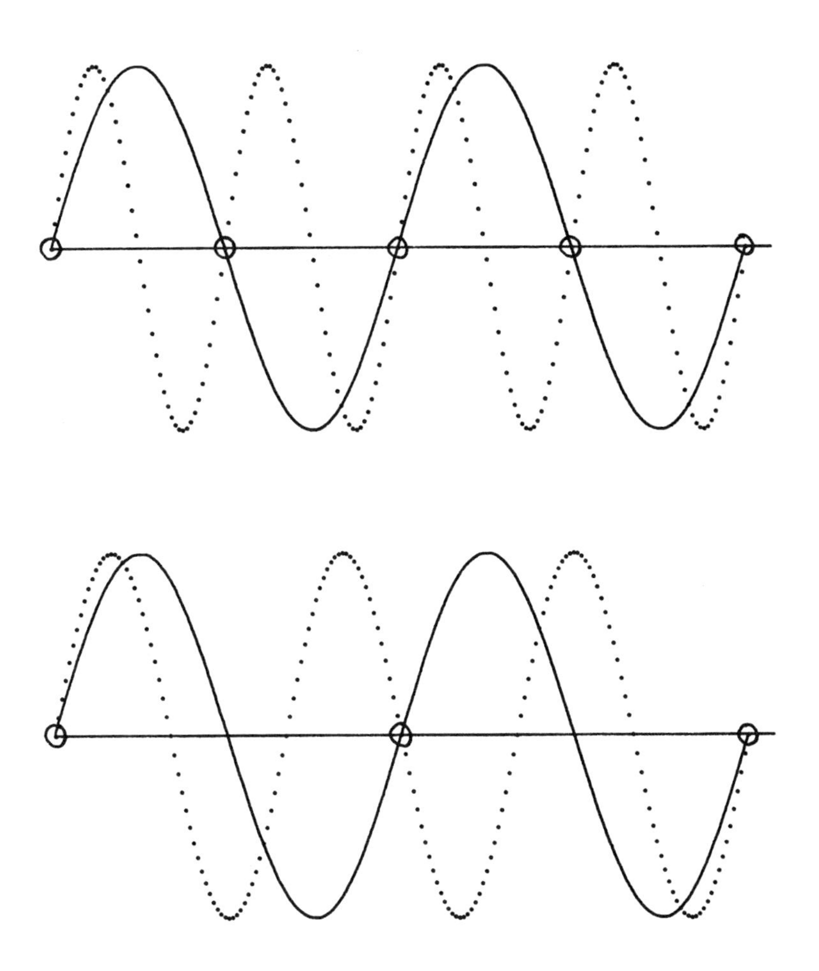

Fig. 30. a) Two notes an octave apart have certain nodes in common; b) two notes a fifth apart have fewer nodes in common. The solid line indicates the lower of the two notes; circles mark the common nodes.

Table 6.

Two Notes a Fifth Apart

1		2	3	4		5	6	7		8	9	. . .
	3/2		3		9/2		6		15/2		9	. . .

In this example, 3/2 means that a note a fifth above 1 has a frequency 3/2 as great. Three of the harmonics occur in common

(3, 6, 9), while the introduction of two new notes (frequency ratios 9/2, 15/2) may produce beats. Sounding a C and a D together produces more beats and hence more roughness, as shown in the following table.

Table 7.

The First Two Notes in the Diatonic Scale

1		2		3		4		5		6		7		8	9	. . .
	9/8		9/4		27/8		9/2		45/8		27/4		63/8		9	. . .

One expects considerable dissonance, since only the last note in the sequence is common to both series.

Helmholtz calculated the amount of dissonance present in two tones, one fixed and the other ranging from equal frequency to one octave higher. He took the harmonic structure present in a typical violin tone as the basis of his calculation. Figure 31 shows the result. To a large degree he has succeeded in predicting the relative unpleasantness of two notes sounded together as ex-

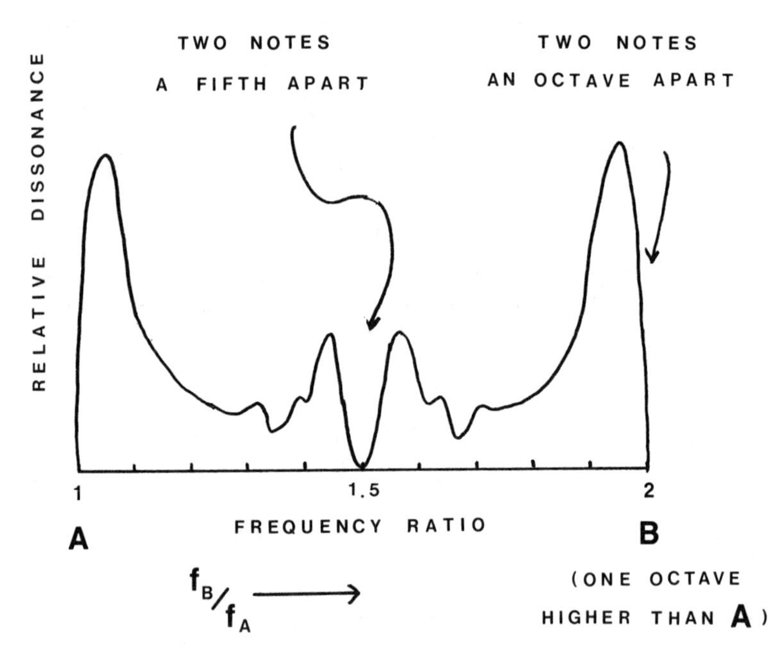

Fig. 31. The relative dissonance between two notes, A and B, as a function of their frequency ratio.

perienced by most people. Many psychological factors enter at this point, and a relative dissonance that sounds unpleasant to one generation of listeners may be acceptable to another generation. A case in point is Mozart's String Quartet in C, nicknamed "The Dissonant," although it is unlikely to strike modern listener's ears as such.

Helmholtz's ideas go a long way toward explaining our perception of consonance and dissonance, but leave unexplained the fact that loud pure tones (sine waves of a single frequency and no harmonics) may give rise to dissonance effects. Modern research shows that the ear has a certain non-linear response, that is, some tissues in the ear do not move an amount proportional to the pressure on them. So a pure tone gives rise to harmonics in the hearing mechanism of the ear. If one sounds two pure tones, one hears extra notes, the sum and difference frequencies of the original notes and their harmonics. The effect becomes more pronounced for louder sounds, and for recorded sounds, because it is very difficult to design a recording instrument with a perfect linear response. As an example, consider two notes of frequency f and F sounded together.

Table 8.

Combination Frequencies

f		2f		3f		4f	.	.	.
F		2F		3F		4F	.	.	.
	f+F		f+2F		f+3F		.	.	.
	f−F		f−2F		f−3F		.	.	.

The analysis is complicated, but just as in the earlier Helmholtz theory, the farther the frequency ratios depart from small whole numbers, the more numerous the sum and difference tones.

C.A. Taylor derives the diatonic scale using only two facts: a musical instrument sounds not only the fundamental but also its harmonics, and an interval of a fifth is the most consonant exclusive of an octave; for these reasons the harmony of this scale is more pleasant than other scales which have been devised. The derivation proceeds as follows: to the base note and its harmonics add a note a fifth above and a fifth below. The three fundamental frequencies are musically consonant. Some of the notes do not lie within one octave (between frequency ratios 1 to 2). Taylor lowers the high notes in octave steps (i.e., multiples them by ½) until they all lie within one octave. Similarly, he raises low notes in octave steps (i.e., multiples them by 2). (See Table 9). When he has reduced all the notes from the three fundamentals and their first six harmonics to one octave, they form the eight notes of the diatonic scale (Table 10). Note that the seventh harmonic does not fit this scheme. Its derivatives are not part of the diatonic scale, so piano makers position the hammers to eliminate the seventh harmonic from vibrations of piano strings.

Table 9.

Derivation of Diatonic Scale

	Funda-mental	2nd Har-monic	3rd Har-monic	4th Har-monic	5th Har-monic	6th Har-monic	7th Har-monic
a) Base Note	1	2	3	4	5	6	7
b) Fifth below base note	2/3	4/3	6/3	8/3	10/3	4	14/3
c) Fifth above base note	3/2	6/2	9/2	12/2	15/2	9	21/2
d) Reduction of a) to one octave	1	1	3/2	1	5/4	3/2	7/4
e) Reduction of b) to one octave	4/3	4/3	2	4/3	5/3	1	7/6
f) Reduction of c) to one octave	3/2	3/2	9/8	3/2	15/8	9/8	21/16

Table 10.

The Diatonic Scale

The reduced fundamental and second through fifth harmonics from Table 9 (lines d, e, and f) are used in making this scale. The sixth harmonic adds no new notes. The seventh harmonic does not contribute to this scheme.

Frequency (in Hz):	264	297	330	352	396	440	495	528
Frequency ratio:	1	9/8	5/4	4/3	3/2	5/3	15/8	2
Note:	C	D	E	F	G	A	B	C_1

The use of the diatonic scale with fixed note instruments such as the piano is impractical. If one wanted to play the same tune starting at D instead of C, the sequence of frequencies in the diatonic scale listed in Table 10 would become the following:

9/8 81/64 45/32 3/2 27/16 15/8 135/64 9/4.

(One obtains this by multiplying the first member of the series, 9/8, by the frequency ratios of the diatonic scale, 9/8, 5/4, 4/3, etc., in turn.) Four new notes appear in this series, and one needs additional notes if the same ascending sequence of notes is to be played starting on the other members of the diatonic series. If one desires to play one instrument in different keys one must modify or temper the diatonic scale.

One such modification is the system of *equal temperament* (Fig. 32) in which one divides the octave into twelve parts so that each note of the rising sequence is in a fixed ratio, K, higher in frequency than its neighbor. Thus, if one begins with a frequency, f, the next higher frequency is Kf, and the next higher one is K^2f, and so on to the last member of the octave, $K^{12}f$, identical with 2f. One can solve the resulting equation, $K^{12} = 2$ on an electronic calculator if it has logarithmic key functions. The answer is $K = 1.0595$. The French mathematician Mersenne correctly determined this frequency ratio and published it in his *Harmonie Universelle* (1636). J.S. Bach demonstrated the usefulness of equal temperament in his 48 preludes and fugues "For the Well-Tempered Clavier" (1722), but the system was not widely put into practice until the nineteenth century.

With equal temperament one may play the same sequence of notes starting on any key, and yet stay within the same pitch

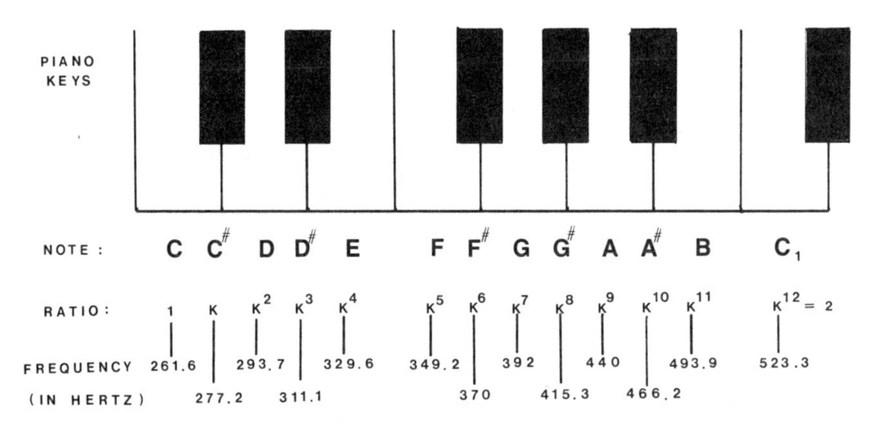

Fig. 32. The system of equal temperament as used on a piano.

range. This flexibility has cost some of the harmoniousness of the diatonic scale. To see what price has been paid, look at the musically important interval, C to G. In the diatonic scale the ratio of frequencies is the harmonious 3/2, while for the equal-tempered scale the ratio is 392/261.6, a difference of 1% from 3/2, or about 0.41 Hz in the middle octave of the piano. The loudness of a piano note has fallen a great deal in two seconds, so the slow vibration of 0.41 per second will hardly be detected. The nature of the compromise results in a slightly increased roughness of chords, a small price to pay for increased versatility. Chords on a piano sound slightly brighter than those on a non-keyboard instrument played in the diatonic scale. One can use this to musical advantage; some piano music sounds dull when transcribed for other instruments.

Tuning a Piano

A piano tuner has a standard-frequency tuning fork (A = 440.00 Hz), and compares other notes with it using a beat-frequency technique. He permits only one string of each note to sound, and damps the others with felt. He compares the second harmonic for A with the third harmonic for D as follows:

A: 440.00 × 2	880.0 Hz
D: 293.7 × 3	881.1 Hz
Difference:	1.1 Hz

He first adjusts the interval for the zero beat condition, then he raises D until he hears about one beat per second. Once he sets D, he compares the fourth harmonic of D with the third harmonic of G as follows:

G: 392.0 × 3	1176.0 Hz
D: 293.7 × 4	1174.8 Hz
Difference:	1.2 Hz

In actual practice he uses a table giving beat frequency rates. In this manner he tunes an entire octave to the equal temperament scale, a process called "laying the bearings." He tunes octaves above and below this in a simple way. To tune an octave above, he listens for zero beats between the second harmonic of the lower note and the fundamental of the higher. Finally he brings all three strings of each note into unison. An assumption in the above analysis is that the overtones of a string are harmonic, but this is not the case, since the stiffness of the strings shifts the overtones slightly higher than a strictly harmonic series as discussed above. Since the tuner starts the tuning process in the middle of the keyboard, this makes the higher notes slightly sharp (higher in frequency than equal temperament), and the lower ones slightly flat (lower in frequency than the equal temperament). This is called "stretched" tuning, and most listeners prefer it to the sound of a piano tuned to a perfectly tempered scale. Instrumentalists (violinists and clarinetists, for example) play sonatas more easily with a piano which is perfectly tempered.

Concert Halls

Singing in the shower is an enjoyable pastime for most people. The "bathroom baritone" is a phenomenon caused by the acoustical properties of a small room with non-absorbing surfaces. First of all, one can build up a large volume of sound with very little effort because of multiple reflections. Consider an isolated sound source compared with one near the junction between a floor and a wall; the sound that would go in all directions in the first case is now concentrated in one-fourth the volume, resulting in four times the loudness. Second, the reverberation time for an initial sound to die away may be two to three seconds. Third, reverberation not only sustains the sound, but changes its quality, attenuating the treble in relation to the bass. A room where this happens to a marked degree is "boomy."

For a concert hall, too little reverberation would not allow the sound level of a given instrument to reach a satisfactory intensity for most listeners. In addition, certain undesirable noise effects of musical instruments (violin bowing noise, the "breathy" sound associated with blowing a flute) occurring primarily in the higher frequencies are not attenuated. Halls with such properties are "dead" or "dry." On the other hand, too long a reverberation time produces unintelligibility and boominess. In general, a reverberation time for music of between one and two seconds is ideal.

Early Christian churches took rectangular Roman basilicas as their models. As they grew larger, speech intelligibility deteriorated, due to echoes and reverberation from the hard surfaces of floor, walls, and ceiling. Speaking gave way to chanting to preserve reasonable intelligibility, and only slower forms of music and singing were possible, as in the Gregorian style. Around 1600, a new type of church architecture evolved in Italy and Sicily. Small chapels or "oratorios" adjacent to larger churches were covered with much sculptural wall detail; for example, the oratorio of San Lorenzo in Palermo, Sicily. The diffusion of sound due to elaborate ornamentation, the moderate size, and the high ceilings of these chapels resulted in acoustics favorable to the development of more complex music. The musical form known as the "oratorio" took its name from the room where St. Philip Neri worked in Rome. Handel's music was often performed at Holywell Music Room, Oxford, a space with acoustics similar to the Italian and Sicilian oratorios.

During the Baroque era, the music of Vivaldi, Haydn, and Mozart was performed in palaces or churches, buildings with stone walls and floors, few draperies, and consequently, long reverberation times. The small orchestras of the time must have sounded much louder in such surroundings than an orchestra of the same size playing in a modern concert hall. Musicologists should be aware that to re-create an authentic version of an older work, they must consider the acoustics of the hall as well as the size of orchestra and type of instruments.

A study of halls specifically constructed for musical performances suggests that their reverberation times ideally suited the favorite music of the period during which they were constructed. In the second half of the nineteenth century, some of the most outstanding concert halls in the world were built in Amsterdam (Concertgebouw), Vienna (Musikverein), and Boston (Symphony Hall). These all have reverberation times intermediate between those of the palaces just discussed and modern halls. These moderate reverberation times suited the sweeter sonorities of the romantic composers such as Brahms, Bruckner, and Wagner. Halls constructed after the Second World War in London (Royal Festival Hall) and Berlin (Philharmonic Hall) have shorter reverberation times, and consequently greater clarity, which is appropriate for the music of Stravinsky and Bartok, for instance.

One may change the reverberation time of a hall by adjusting the ratio of acoustically absorbing surfaces to the volume of the hall. Since the optimum reverberation time for maximum intelligibility of speech is lower than that for music, a hall designed primarily for speech would not have the best characteristics for music. Similar considerations apply to the design of opera houses, which need smaller reverberation times than concert halls. The designer must solve many other problems, such as avoidance of echo and the loss of sound under balconies. Experimentation with full-scale halls is prohibitively expensive, but acoustical engineers have recently developed scaling techniques for use with architectural models. Their use before a hall is constructed should greatly benefit future concert-goers.

Bibliography

Backus, John, *The Acoustical Foundations of Music* (Norton, N.Y., 1969).

Bartholomew, W.T., *Acoustics of Music* (Prentice-Hall, Englewood Cliffs, N.J., 1942).

Benade, A.H., *Horns, Strings and Harmony* (Doubleday Anchor Books, Garden City, 1960).

Beranek, Leo L., *Music, Acoustics and Architecture* (J. Wiley, N.Y. 1962).

Bergeijk, W.A. van, Pierce, J.R., and David, E.E., Jr., *Waves and the Ear* (Doubleday Anchor Books, Garden City, 1960).
Berlioz, Hector, *Treatise on Instrumentation* (re-edited by Richard Strauss) (Kalmus, 1948).
Coltman, John W., "Acoustics of the flute," Physics Today, **21,** 25, November 1968.
Culver, C.A., *Musical Acoustics* (Blakiston Co., Toronto, 1947).
Drake, Stillman, "Renaissance Music and Experimental Science," Jour. Hist. of Ideas, **XXXI,** 483, Oct.-Dec. 1970.
Harris, Cyril M., "Acoustical Design of the John F. Kennedy Center for the Performing Arts," J. of the Acoustical Soc. of Amer., **51,** No. 4 (Part 1), 1113, 1972.
Hutchins, C.M. (ed.), *The Physics of Music* (Readings from *Scientific American*) (W.H. Freeman and Co., San Francisco, 1978).
Hutchins, C.M., and Fielding, F.L., "Acoustical Measurements of Violins," Physics Today, **19,** 35, July 1968.
____. "Founding a Family of Fiddles," Physics Today, **20,** 23, February 1967.
Jeans, Sir James H., *Science and Music* (Macmillan Co., 1953).
Josephs, Jess, *The Physics of Musical Sound* (D. Van Nostrand, Princeton, N.J., 1967).
Lindsay, R.B., "Science and the Humanities," in *The Role of Science of Civilization* (Harper & Row, N.Y., 1963), Chap. 3.
Lloyd, L.S., *Music and Sound* (Oxford U. Press, 1937).
Miller, D.C., *The Science of Musical Sounds* (Macmillan, 1922).
Palisca, Claude V., "Science and Music," in *17th Century Science and the Arts,* H.H. Rhys (Ed.) (Princeton Univ. Press, 1961).
Richardson, E.G., *The Acoustics of Orchestral Instruments* (E. Arnold & Co., 1929).
Rigden, John S., *Physics and the Sound of Music* (John Wiley and Sons, New York, 1977).
Risset, Jean-Claude, and Matthews, Max V., "Analysis of Musical Instrument Tones," Physics Today, **22,** 23, February 1969.
Roederer, Juan G., *Introduction to the Physics and Psychophysics of Music,* 2nd Ed. (Springer-Verlag, New York, 1975).
Shankland, Robert S., "The Development of Architectural Acoustics," American Scientist, **60,** 1972, March-April, p. 201.
Strong, W.J., and Plitnik, G.R., *Music, Speech and High Fidelity* (Brigham Young University, Provo, 1977).
Taylor, Charles A., *The Physics of Musical Sounds* (American Elsevier Publ. Co., N.Y., 1965). Includes a recording of acoustical illustrations (45 rpm).
Wood, Alexander, revised by Bowsher, J.M., *Physics of Music* (Methuen, London, 1964).

Chapter 7

Architecture and the Dome of Heaven

TO CHRONICLE THE myriad influences that science has had on architecture would take a large volume. The purpose of this chapter is more limited: to spotlight instances where architecture and science have interacted directly. A particularly clear example occurs when an architect attempts to design a building used for a scientific purpose, such as an observatory. The laws of physics impose some constraints on architectural imagination: a building housing an optical telescope must have viewports, to cite an obvious example. But within rather broad limits the architect is free to design practical working environments expressing the spirit of adventure in scientific investigation.

From earliest times architecture has had an intimate relationship with astronomy. One suspects that many of the monuments of pre-history, such as the ziggurats of Babylon, the stone and wood henges of Britain, and the pyramids of Mexico and Egypt, may have been used for astronomical observations. The first architect whose name we know, Imhotep, built the stepped pyramid at Saqqara for King Zoser (Third Dynasty of Egypt). Rather than indulge in speculation about the purposes of these ancient structures, fascinating though that may be, I will begin in seventeenth-century Europe at the time of Isaac Newton when scientific societies were forming and beginning to establish programs of astronomical research. At that time two famous scientist-architects at the behest of their respective kings designed the great observatories at Paris (1667) and Greenwich (1675). In spite of their parallel beginnings, the outcomes were quite different.

The Paris Observatory

Claude Perrault. How did one become an architect at a time when architecture was not an established field in which one could be apprenticed or for which one could study? It helps to have a younger brother with connections at court as we shall see below. Trained as a physician, Claude Perrault worked primarily in the fields now known as comparative anatomy, physiology, and botany. In 1666 he became a founding member of the *Acadêmie des Sciences*. Perrault and a group of anatomists from the *Acadêmie* undertook the task of dissection of a series of mammals, birds, reptiles, amphibians, and fish over a period of years, whenever an animal in the kings's collection would die and thus become available for examination. While investigating such legends as the ability of salamanders to live in fire and the ability of chameleons to live on air, Perrault made some real discoveries, learning, for example, that the chameleon could protrude its tongue and swivel each eye independently.

Fig. 33. The Paris observatory. Note that several of the optical instruments had to be used in the open courtyard rather than inside the building.

An intrigue at the court of Louis XIV preceded Claude's first architectural splash. Claude had a younger brother, Charles, who was the author of many of the Mother Goose stories. Charles was a friend and confidant of Jean Baptiste Colbert, financial advisor and superintendent of buildings for the king. When an addition to the Louvre was needed, Colbert (1664) invited French architects to submit plans, and Charles advocated his brother's plan. But the king, having other ideas, invited the celebrated Italian architect Giovanni Bernini to come to Paris and draw up plans for the Louvre. The first stones of the new Louvre were laid in 1665 according to the Bernini plan, but Bernini quarreled with Colbert and left Paris indignantly. After Bernini's departure in 1666, Claude intimated to the king that Bernini had intended to tear down some of the older parts of the Louvre. On this pretext Colbert ordered new plans drawn up. In 1667, Louis LeVau and Perrault submitted their designs to the king, who chose Perrault's scheme. One may view the Perrault facade today on the east side of the Louvre. The Bernini plan is also extant, and the reader may judge who had the better design.

Simultaneously with his work on the Louvre (1667) Perrault worked on the building which Colbert conceived of as an observatory, museum, and meeting place for the *Acadêmie*. From the beginning the multiple purposes doomed it as an effective observatory. Perrault's idea for the building was a rectangular block with octagonal towers at two corners, southeast and southwest, and a square tower in the middle of the north side. The windows were tall and rounded at the top. Despite the balustrade surmounting the flat roof and the sculpted sides of the central south window, the design as a whole is rather plain. The building is still standing and may be seen today when one visits the *Musée Astronomique de l'Observatoire de Paris.*

The Cassini "Dynasty" at Paris. Colbert invited the Italian astronomer Gian Domenico Cassini (Cassini I) to visit Paris for a limited period to help set up the observatory. He found French life so agreeable that he eventually married a French woman, obtained French citizenship, and became director of the observatory. Thus began an association which for over a century made the family Cassini and the Paris Observatory virtually synonymous. For his astronomical work Cassini wanted to install quadrants on the four walls of the central room of the second story, but attached towers blocked the view. To get around this problem, he proposed terminating the towers at the first story. In addition, he wanted the central observing room of the second story surrounded by an unroofed corridor. Perrault refused to listen to the astronomer's advice or to change his design. As a concession, however, he did leave the southeast tower and the central salon unroofed. Although Cassini was able to get a large official subsidy for the purchase of instruments, the new telescopes often had to be used out in the garden rather than in the building.

In spite of the difficulties posed by the observatory building, Cassini I, an excellent observer of the heavens, discovered four new satellites of Saturn (Iapetus, Rhea, Tethys, and Dione). In 1675, he observed that the ring surrounding Saturn was divided into two parts separated by a narrow band now known as *Cassini's division*. But he was more successful in observations than in

Fig. 34. Louis XIV visiting a meeting of the Royal Academy of Sciences in the King's Library. The central figures are Louis XIV and Colbert. To the left of these two are Louis de Bourbon and the King's brother. Between them, talking over his shoulder and pointing, is Gian Domenico Cassini. Claude Perrault stands just behind Colbert and the King. The Paris Observatory, then under construction, is visible through the right-hand window. Engraving by Sebastien Le Clerc (1671).

theorizing. Whether through conservatism, fatalism, or sheer stubbornness he had a knack for picking the losing side in scientific controversy. He was a confirmed Cartesian and an opponent of the theory of universal gravitation and of Römer's ideas of a finite speed of light to explain certain irregularities in Jupiter's satellites.

His son Jacques (Cassini II), a staunch defender of his father's ideas and another anti-Newtonian, succeeded him. In one of his books he mustered arguments defending the elongation of the terrestrial ellipsoid, in contrast to the school of Newton and Huyghens which supported the idea that it was flattened along the line joining the two poles.

To complete guidelines for a new map of France and to check on the shape of the earth, his son Cassini de Thury (Cassini III) launched new geodesic expeditions. The results of one expedition seemed to favor strongly the Newtonian hypothesis of flattening at the poles, so Cassini III had the courage to reverse his position from that held previously by his family. But the matter was not competely laid to rest until the English astronomer James Bradley noticed a small wiggle in the earth's motion which he explained as the effect of the gravitational attraction of the sun and moon on the excess matter at the equator. Cassini III is remembered more for his cartographic than for his astronomical work: he produced the first map of France according to modern principles.

His son Jean-Dominique (Cassini IV) was born at the observatory and continued his father's work on the great map of France. He also participated in the survey to link the Greenwich and Paris meridians. He opposed the French Revolution, and the authorities forced him to leave the observatory on September 6, 1793, thus ending a period of 120 years in which the family had controlled the Paris Observatory. His son had no male heirs, so the Cassini name was extinguished in France.

The Royal Observatory at Greenwich

Sir Chrisopher Wren. Inscribed on a plaque near Wren's tomb in St. Paul's Cathedral are the Latin words, *"Lector, si monumentum requiris, circumspice,"* (Reader, if you seek a monument, look around). Indeed Wren needs no other monuments than St. Paul's and the City of London churches rebuilt after the fire. Many of the churches have survived the World War II bombardment or have been rebuilt. Wren's churches reflect the influence of classical style, especially that of the Roman architect Vitruvius. But he did more than copy the Roman version; he adapted, simplified, and injected his own original conceptions, as in the City churches where the old foundations, perhaps cramped between two earlier buildings or occupying an odd-shaped lot, had to be used. A purely classical building scheme simply wouldn't fit without a great deal of ingenuity. The City church towers are some of his most inspired creations. Take, for example, the tower of St. Mary-le-Bow (1680), which rises squarely from the ground. Above the bell stage is a circular classical temple topped by a ring of flying buttresses (bows). The buttresses support a smaller square temple which in turn is surmounted by an obelisk. Thus Wren achieved the classical equivalent of the Gothic steeple. Since no Roman spire or tower appears in Vitruvius' books, in this project he had to be guided by his own artistic instincts.

Followers of Wren imitated his mannner throughout the world. His pupil, James Gibbs, designed the famous St. Martin's-in-the-Fields whose spire has become identified as the symbol of the Anglican Church. Wren's influence has made itself manifest in the United States in Christ Church, Boston (1723), and in the dome of the Capitol in Washington, D.C. (1855-65). In France, the church of Ste. Geneviêve (now the Panthéon) has a collannaded drum and dome whose profile reflects that of St. Paul's. Similarly, the dome of St. Isaac's (1840-1842) in Leningrad could be added to the list.

Wren began his architectural endeavors after he was thirty. Before that he was a distinguished member of the Royal Society and Savilian Professor of Astronomy at Oxford. The range of his accomplishments is staggering. Newton in the second edition of *Principia* (1713) called him one of the leading geometers of the day, along with John Wallis and Christiaan Huyghens. The Royal Society had asked the three of them to report their researches into the laws of impact between solid bodies. Wren was also well versed in pure mathematics; he solved the problem of the rectification of the cycloid, that is, finding a straight line with the same length as the cycloid. (The cycloid is a curve formed by the motion of a point on a circle's circumference as the circle rolls along a straight line). In addition to astronomy, physics, and mathematics, he was also active in the medical area, making a series of drawings of the dissection of a human brain for the anatomist Thomas Willis. He was a skillful model maker, whether of Saturn's rings, the action of muscles, the mountains on the moon, or the eye. Robert Boyle and Wren made the first injection of a foreign substance directly into an animal's bloodstream. For this experiment they used a large dog, and exposed one of the veins in the hind leg. Puncturing the vein with a quill, Wren introduced a tincture of opium, and observed the effects of the drug almost immediately, a milestone in surgical history. Wren followed this up with trials of other drugs, culminating in 1659 with the transfusion of blood from one animal directly into another.

Puzzling to many modern readers is Wren's transition from astronomer and mathematician to architect. One must realize that, in seventeenth-century England, architecture (subdivided into civil, naval, and military) was viewed as a branch of mathematical science. Military architecture consisted not only of fortifications, but gunnery and surveying, highly mathematical occupations. Furthermore, the Royal Society maintained an interest in architecture, receiving Wren's design for a surveying level, for instance. Actually, in his circle Wren is exceptional not in taking up architecture but in succeeding at it. Robert Hooke, later a famous microscopist, also dabbled in architecture in his position as City Surveyor, assisting in the rebuilding of London after the fire and often discussing St. Paul's with Wren. Wren's abilities as a modelmaker and draftsman were well known to the king. Significantly, his first two commissions were for advice concerning the remodeling of existing structures, the harbor at Tangier (Wren rejected this) and the strengthening of old St. Paul's, tasks which could logically have been given either to a scientist or an architect. Since building activity had been at a low ebb during the Commonwealth era, Wren's ability could not have shown itself earlier. His great opportunity came with the Great Plague and Fire. During the plague the Royal Society did not meet, so Wren went to Paris to see the state of architecture there, viewing construction in progress at the Louvre and briefly meeting Bernini. When the Great Fire broke out shortly after his return from Paris, Wren was ready with a plan for reconstruction of the City even before the embers were cool. His design, a magnificent one, was never realized because many buildings were rebuilt on their old foundations, but he did transform the City skyline with his famous steeples.

One of Wren's few secular buildings is a structure designed for scientific research, the Royal Observatory (1675). When King Charles II founded the observatory, the oldest scientific institution in Great Britain, he selected a site at his park in Greenwich and asked the Surveyor-General, Christopher Wren, to produce a suitable design. Since Wren must have taken a personal interest in the project because of his former work in astronomy, why isn't the result characteristic of most of Wren's architectural work? The reasons for this are two-fold. First, little money was available for this project: evidence for this is that it was built with old materials from the Tower of London plus surplus bricks from Tilbury Fort. In addition, the design had to conform to the building's use as an observatory, and ornaments which might interfere were ruled out. The result is an un-Italian, fortress-like brick building situated on a hill with a commanding view of the River Thames. The structure has two stories linked by large scrolls: the lower story

Fig. 35. A view of the north side of the Royal Greenwich Observatory. Etching by Francis Place.

Fig. 36. This magnificent Wren interior, the Octagon Room at Greenwich, has survived almost intact to this day. Above the door are portraits of King Charles II and James, Duke of York. On the left are two of Tompion's clocks with 13-foot pendulums hidden behind the wainscot. The pendulum bobs can be seen in the windows above the clocks. Etching by Francis Place.

consists of living quarters for the Astronomer Royal, and the upper is a large octagonal room for general observations and occasions of state. In Wren's words the building was intended "for the observator's habitation and a little for pompe."

The original observatory designed by Wren, together with latter additions, is open to the public as part of the National Maritime Museum. Down the hill from the observatory are the main museum buildings, including the Queen's House (1616-1635) designed by Inigo Jones. Between the Queen's House and the river stands Wren's magnificent Greenwich Hospital (now the Royal Naval College), the last and finest of his secular buildings. All of the above mentioned buildings constitute one of the most distinguished architectural groups in the country.

The Royal Astronomers at Greenwich. The history of the Royal Observatory is intimately related to the search for a practical method of determining longitude at sea. Latitude is simply determined by observing the elevation of the sun at midday. It would be equally easy to determine longitude if the moment of midday in an unknown position could be compared with the time as registered at Greenwich. Each hour of difference corresponds to 15 deg. of longitude. Now the time signals are broadcast by radio, but no comparable service was available in Wren's age. Reliable marine chronometers were not built until the middle of the eighteenth century. Galileo's suggestion that the moons of Jupiter be used as a clock was not practical, due to the difficulty in making a table of predicted positions and the difficulty of observation from the deck of a ship. Newton pointed out that observations of the moon's position should be easier to make, and he tried without success to work out tables of the future positions of the moon. It wasn't until Euler's mathematics made possible accurate calculations that such tables of the moon's position could be published. The first Astronomer Royal, John Flamsteed (1646-1719), assisted the moon project by measuring the bearings of nearly 3,000 stars to within an accuracy of 10 seconds of arc (about ten times more accurate than Tycho Brahe's data, see Chap. 2), thus providing the foundation for modern positional astronomy. As it turned out, John Harrison invented an accurate timekeeper at about the same time that Tobias Mayer published moon tables. But the marine chronometer was slightly more accurate; it gave longitude to within 1 minute of arc, while the tables gave results to within 4 minutes of arc.

Although he is better known for computing the orbit of the comet which bears his name, the second Astronomer Royal, Edmond Halley (1656-1742), also invented a method for determining the scale of the solar system. In rare instances when the Sun, Venus, and Earth lie along a straight line, Venus appears as a dark spot moving across the sun's face. The path Venus appears to take across the sun's surface differs for observers in different geographical locations. Halley arranged for observations of transits of Venus in 1761 and in 1769. In the latter year, Sir Joseph Banks and Captain Cook made observations from Tahiti, establishing the absolute scale for the solar system to within an accuracy of 5 per cent. The National Maritime Museum in Greenwich possesses a chronometer used by Captain Cook on his second and third voyages to the Pacific and also one used by Captain Bligh and stolen by the mutineers aboard the *Bounty*. These

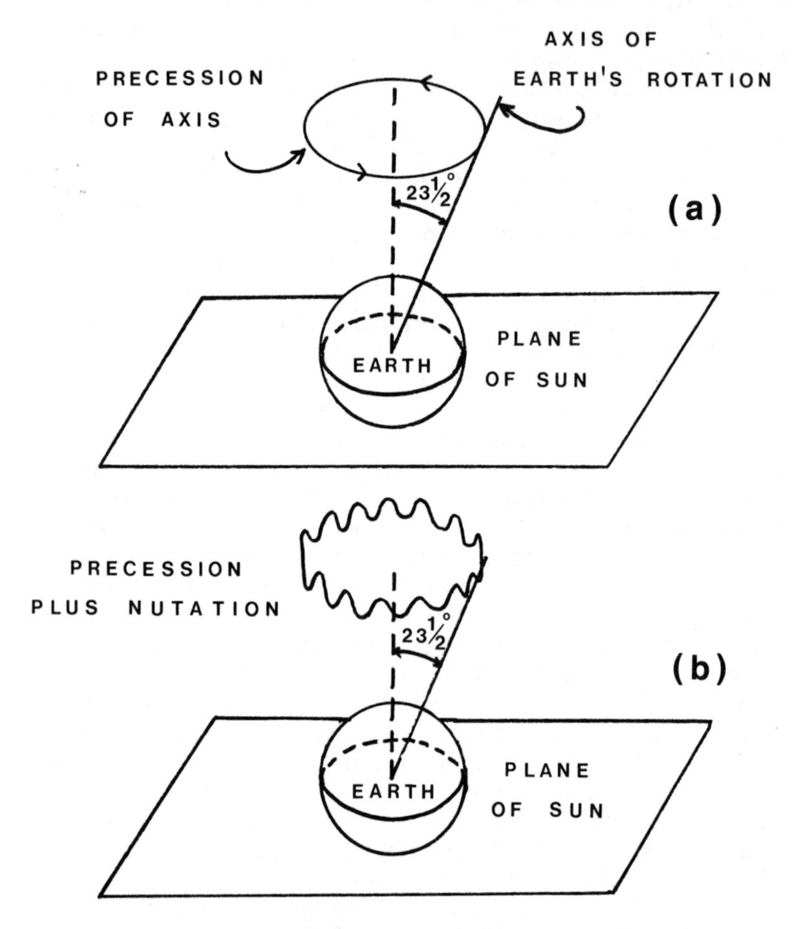

Fig. 37. a) The rotation axis of the earth precesses about the dashed line just as a top precesses about the vertical direction. A complete revolution takes about 26,000 years. b) James Bradley discovered fluctuations (nutation) with a period of 18.6 years in the rate of precession.

instruments may be found in the Navigation Room along with a series of large and beautiful chronometers designed by John Harrison.

Halley's successor, James Bradley (1693-1762), made two discoveries while pursuing accuracy in measuring star positions. The first is the aberration of starlight, an apparent shift in direction of starlight as it approaches earth due to the combined effect of the motions of the earth and light. As a result, a star appears displaced from the true position which it would have in absence of the earth's motion. This effect varies with the time of the year and with the angle the starlight makes with the plane of the earth's orbit, but the order of magnitude is about 20 seconds of arc. The second discovery of Bradley's (referred to in the discussion of the Cassinis) was a small effect called *nutation* which is superimposed on a larger motion called precession. The precessional motion of the earth's polar axis is like the motion of the spin axis of a child's top but with a period of some 26,000 years. The effect is quite small, only about 20 seconds of arc per year, but Bradley could measure changes even smaller than that. He discovered that the precession had a kind of "wiggle" (nutation) whose period was only 18.6 years. Precession and nutation are caused by gravitational pulls which the sun and moon exert on the nonspherical earth.

The fourth Astronomer Royal, Nathaniel Bliss, died within two years of his appointment and so did not have time to leave his mark on Greenwich. The next Astronomer Royal, Nevil Maskelyne (1732-1811), published the first *Nautical Almanac* in 1767, and it has been published every year since. Maskelyne also measured the apparent motion of the stars which arises because of the real motion of the entire solar system through space. The next Astronomer Royal, John Pond, introduced the Time Ball, a leather-covered wooden framework about 5 feet in diameter dropped from a 15-foot mast on top of the observatory at precisely one o'clock. It proved to be so popular and useful to coastal navigation that several were installed in other locations and linked to Greenwich by telegraph lines.

The last Astronomer Royal to be appointed under Royal Warrant was George Biddell Airy (1801-1892). He conducted stellar aberration experiments with a water-filled telescope, part of the experimental background which preceded Einstein's Theory of Special Relativity. He used a combination of spherical and cylindrical lenses to correct astigmatism in 1825, but it was not until 1862 that ophthalmologists began to adopt this method on a large scale. He was also the first to calculate the diffraction pattern of a circular opening (a pattern of circular dark fringes caused by interference of light waves), thus establishing that there were ultimate limitations to the sharpness of lens-produced images even if scientists solved all of the problems of glass-grinding and figuring.

After the retirement of Airy, an international conference at Washington, D.C. in 1884 chose Greenwich as the Prime Meridian. As a result, scientists measure all lines of longitude relative to that passing through Greenwich, and they refer to time there as Greenwich Mean Time. But the glory of Greenwich's earlier days faded as the spotlight turned to the construction of large-diameter reflecting telescopes in the United States. In 1946 the government gave the observatory a new home at Herstmonceux Castle in Sussex, away from the smog of London.

Observatory Designs at Paris and Greenwich Compared

The observatories at Paris and Greenwich quickly became astronomical monuments enveloped in an aura provided partly by the role they played in significant astronomical advances and partly by their architectural roots. Although the various Royal Astronomers at Greenwich complained about the lack of instruments, the basic design of the building was sound for at least two reasons. First, Wren had experience as an astronomer, and thus knew intimately the requirements for a functioning observatory. Also the warrant of Charles II founding the Royal Greenwich Observatory was firmly grounded in a practical purpose:" . . .in order to the finding out of the longitude of places for perfecting navigation and astronomy . . ." In contrast the Paris Observatory was well equipped with instruments, but the building design hampered astronomical work. Perrault's work in botany and anatomy did little to prepare him for the technically demanding task of designing an observatory. By the time the astronomer Cassini I arrived on the scene, the first story had already been completed. Colbert's intention was as much to add glory to Louis XIV as to further astronomical studies; probably no such multiple-purpose building could have satisfied the astronomers.

Meanwhile, another optical development, compact reflecting telescopes made by Isaac Newton and Robert Hooke, eased observatory space requirements. The reflectors with their folded optical paths could achieve the same magnification in less distance than their refracting counterparts. In other words, a telescope utilizing mirrors is generally shorter than a telescope utilizing lenses to achieve the same magnification. To shelter such a reflector required only a lightweight dome. But it took a long time for reflectors to come into general use; the first of the now-familiar domed buildings, George III's Richmond Observatory (1769), rose about a century after Greenwich Observatory. Meanwhile an astronomical observatory constructed in India forty

Fig. 38. In left and right foreground the Samrat Yantra of Delhi, in the background a composite instrument.

years after Greenwich took a strikingly different form.

The Observatories of Jai Singh

In contrast to the ornate temples from India's past, the Delhi Observatory or Jantar Mantar has the stark geometric simplicity of the monoliths of Stonehenge or today's skyscrapers. In fact, the three principal masonry buildings date from the first quarter of the eighteenth century.

Samrat Yantra. The largest and most imposing of the observatory structures is nothing more than an enormous sundial. It consists of a triangular gnomon with hypotenuse parallel to the earth's axis and on either side a circular quadrant parallel to the equator. The shadow of the gnomon on the quadrants gives the local time. The gnomon at Delhi is 68 feet high, surpassed only by the 90-foot one at Jaipur. A staircase ascends the gnomon; it was originally in marble, but is now coated with lime plaster.

Jai Prakas. This instrument is a concave hemisphere representing the heavens. Part of the surface of the bowl is cut away leaving "teeth" which are graduated in degrees. Because of the access provided by the "teeth," an observer may enter the instrument, place an eye at a graduated line, and observe the passage of a heavenly body across the intersection of two crosswires stretched across the mouth of the bowl. At Delhi duplicate instruments 27.5 feet across were built so that the "teeth" of one instrument complement the openings of the other. The two instruments together are equivalent to one instrument whose surface is completely covered with gradations.

Ram Yantra. A large unroofed cylindrical building with central pillar constitutes this instrument. Many arched openings forate the outer wall. The central pillar is of diameter such that the distance from the circumference of the pillar to the wall (24 feet 6½ inches at Delhi) is equal to the height of the wall. The floor is broken into angular sectors of width six degrees. Similarly the walls are graduated and have notches for placing sighting bars. By sighting along a bar one may read off the horizontal (azimuth) and vertical (altitude) angles of the position of a heavenly body. Like the Jai Prakas, this instrument is also duplicated, so that a column in one is an opening in the other.

What sort of man devised these monumental instruments? The Maharaja Sawai Jai Singh II of Jaipur was born in 1686, one year before the publication of Newton's *Principia*. In the midst of anarchy and warfare he managed to establish a new capital, Jaipur, named after him. He found the Hindu calendar and astronomical tables inaccurate, and conceived of a program of astronomical studies which occupied his whole life. To these ends, he constructed not one but five observatories in Delhi, Jaipur, Ujjain, Benares and Mathura. He studied Hindu, Moslem, and European methods, but the Moslem astronomer Ulugh Beg influenced him most.

His indifference to European achievements in planetary theory and instrumentation is remarkable. He must have known about Brahe, Kepler, and Galileo through the Flamsteed book, *Historia Coelestis Britannica,* which he possessed. Perhaps these scientists did not make a bigger impression on Jai Singh because his principal European contacts were with Roman Catholic priests. Since Galileo's books were not taken off the *Index* until 1835, the priests were hardly in a position to advocate Copernican and Galilean teachings. Greenwich Observatory was founded some forty years before the one at Delhi, and its instruments could have served as models. But Jai Singh turned his back on the many European innovations in astronomical instruments made during the seventeenth century, for example, the telescope, micrometer, vernier, and tangent screw. He believed that the lighter metal instruments would suffer from vibration, bearing wear, and misalignment; furthermore, he felt that their gradations were too small for accurate measurement. In his view, large masonry buildings were preferable. But the very permanence of his designs hindered further improvements. G.R. Kaye claims that a single modern theodolite is worth more as an astronomical instrument than all of Jai Singh's buildings. He made no new astronomical discoveries, but he did achieve what he set out to: rectification of the calendar and prediction of eclipses. In 1743, his wives and concubines expired with him on the funeral pyre, leaving behind his colossal astronomical instruments.

A Cenotaph for Newton

As we have seen in the history of the Cassini family, the theories of Newton made slow headway in France, where the ideas of Descartes held sway. In 1699, Newton had been elected a Foreign Associate of the *Acadēmie,* but for years did not communicate with that body. The publication in 1722 of the Paris edition of the French translation of *Opticks,* promoted by several powerful figures including the Chancellor of France, marked a turning point in the acceptance of Newtonian ideas in France. Most French scientists had to acknowledge the reality of Newton's experiments in light and color, despite whatever reservations they had with regard to the interpretation of those experiments. Admitting that *Opticks* was an important work, they moved one step closer to the mathematically arduous *Principia.* By the end of the eighteenth century the triumph of Newtonian mechanics was all but complete, ending the French hostility to Newtonianism but not the scientific rivalry between the two countries.

In the 1780's the *Acadēmie Royal d'Architecture* thought it appropriate to propose a contest on the theme of "a tomb in honor of Newton. This monument, dedicated to the glory of the great genius, ought not to be magnificent so much as imposing in its dignified grandeur and simplicity." Several of the designs submitted to the academy are extant, among which are those of Delēpine and Gay. But the most striking design is that of Etienne-Louis Boullēe (1784). He wrote of Newton in a rapurous vein, "Sublime mind! Vast and profound genius! Divine Being! Newton! Accept the homage of my weak talents . . . O Newton! . . . I conceived the idea of surrounding thee with thy discovery, and thus, somehow, surrounding thee with thyself." His conception is a hollow sphere of geometric simplicity and colossal proportions set in a cylindrical base much like an egg set in a gigantic eggcup. He imagined the terraced base planted with trees, producing an effect not unlike an imperial Roman tomb such as *Castel San Angelo.* Inside the immense sphere Boullēe wanted only a sarcophagus on a raised catafalque. To a large extent he captured the essential spirit of Newton. The weakness of the design, apart from its impracticality, is a certain theatrical quality: he envisioned holes in the top of the sphere which would simulate

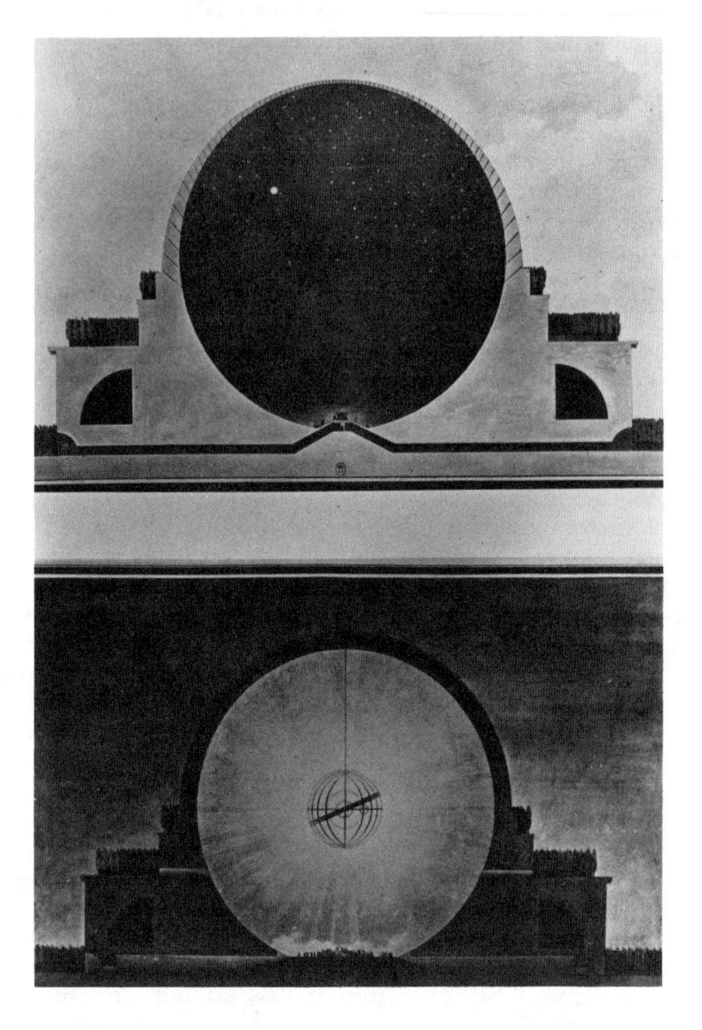

Fig. 39. Cross-section of Newton's cenotaph. Top: interior night effect. Bottom; interior day effect.

stars and planets when illuminated by daylight, while at night light would be supplied from a fixture within an armillary sphere suspended at the center of the globe.

Boullée's cenotaph was an academic exercise, and was never intended to be actually constructed. In concept it is similar to a present-day monument to technology, *Evoluon,* the large exhibition building of the Philips electronic industries in Eindhoven, Holland. Although some critics dismiss the design as "world's fair architecture," the spheroidal shape is not only practical, but

Fig. 40. Eric Mendelsohn's Einstein Tower.

suggests a flying saucer or atom. Such associations complement the scientific exhibits inside.

The Einstein Tower

In the trenches during World War I, a young man made sketches of imaginary structures much as Boullée had done. These imaginative, fantastic structures Eric Mendelsohn called prosaically, "Architecture in Steel and Concrete." In his drawings he envisioned huge spans, cantilevers, and sculptural forms. At first he did not have the know-how to construct such free-flowing forms; he had to turn to engineers for the solution of these problems. With the accumulation of experience he learned how to turn his dreams into reality. Today, with the aid of steel and reinforced concrete, many buildings have shapes reminiscent of Mendelsohnian fantasies; Eero Saarinen's TWA terminal at John F. Kennedy airport and Jorn Utzon's Sidney Opera House are two examples.

After the war, his first commission was for an observatory and laboratory in Potsdam to test certain predictions of Einstein's

General Relativity Theory. The astrophysicist E.F. Freundlich, a friend of Mendelsohn's, designed an observatory which projected the image of the sun or a star to an underground constant-temperature room. A dispersing element then split the light into its spectral components, and the predicted gravitational "red shift"was sought. Both public and private sources contributed money to this project. Mendelsohn later said that his design sprang from "the mystique around Einstein's universe." His imaginative yet practical creation came to be known as the Einstein Tower (1919). The flowing curves in reinforced concrete expressed the intellectual excitement which accompanied Einstein's theories, while the observational dome was organically joined to the laboratories underneath so that the function of the building was clearly intelligible. This building created a sensation in Germany and successfully launched Mendelsohn into his career.

The Kitt Peak Observatory

Looming on top of a mountain in Arizona like a giant white shoe is the solar telescope housing designed by Myron Goldsmith of the architectural firm of Skidmore, Owings, and Merrill for the Kitt Peak National Observatory. The "heel" is a vertical tower 100 feet high, while the "instep" is a 500-foot diagonal tunnel of which only two-fifths is visible above ground.

Since the path of the sun is restricted to a small portion of the sky, a telescope designed for solar work need not point to all parts of the celestial hemisphere. The designers chose a fixed diagonal shaft aimed at the north celestial pole and topped by a sun-tracking mirror or *heliostat.* In thus sacrificing mobility the designers gained structural rigidity for what is in effect a camera with a 300-foot focal length. The exceptionally long focal length enables the users to obtain an image of the sun 33.5 inches in diameter. The architects had to surmount two major hurdles in the design: vibration caused by wind and optical distortion caused by thermally induced air currents. To solve the first problem they devised a windshield of copper sheeting structurally separated from the concrete tower and optical tunnel. The effort to create thermal equilibrium took several different approaches. They painted the windshield white to reflect heat; in addition, they bonded tubes carrying cooling liquid to the windshield to stabilize its temperature. To get above the layer of unevenly heated air near the surface, they located the heliostat 100 feet above ground. The interior of the diagonal shaft can also be cooled when the temperature of the surrounding rocks is higher than that of the air. If provision for cooling had not been made, the shaft under these circumstances would act like a giant chimney.

Fig. 41. McMath Solar Telescope, Kitt Peak National Observatory.

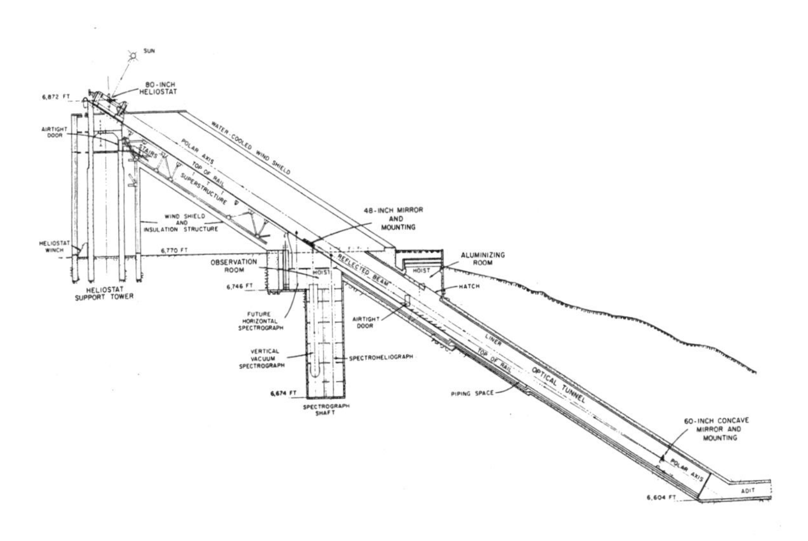

Fig. 42. Outline drawing showing the light path in the McMath Solar Telescope.

The result is a primal shape which recalls Jai Singh's structures. In both cases the shapes clearly reveal the purpose for which they were built.

Conclusions and Summary

Three of the observatories we have considered (Paris, Greenwich and Delhi) were designed by scientist-architects, a circumstance which may no longer be possible given the explosive growth of science and the resultant specialization necessary for individuals seeking to master portions of that vast amount of knowledge. Earlier in this century, architectural training included a study of calculus and physics; now these subjects are no longer required in most architectural schools, so modern architects must consult engineering firms for technical advice. A remnant of the old ways persists in the pleasure which most architects take in designing furniture. Mies van der Rohe, Marcel Breuer and Eero Saarinen didn't need to call in a firm of structural engineers to see if their chairs would support a human body! Happily a few still retain the old tradition of scientist-architect, for example Buckminster Fuller and his geodesic dome.

Although constrained by the laws of optics for good viewing, each observatory participated fully in the architectural tradition of its time. The Paris and Greenwich observatories proudly bear the classical imprint. Few modern observers looking at these observatories could guess their function even though they have fewer ornamental elements than most royally commissioned buildings of their time. At Delhi and Kitt Peak the architects dispensed with buildings in the usual sense because the instruments themselves are so large that scientists could work inside them. The astronomical functions of these "buildings" are unmistakable. Kitt Peak is an outstanding example of the "brutalist" school of architecture; it also functions well as a solar observatory. In this case form and function, each excellent in its own right, have blended into a superb whole. The radio telescope filling an entire mountain valley in Arecibo, Puerto Rico, recalls the "earth sculpture" of Christo and other artists in which large portions of landscape are plowed, painted, wrapped, or fenced.

In addition to housing a scientific instrument with attendant staff and reflecting architectural tradition, each observatory embodied an idea. To assert that an observatory building can come to symbolize an intellectual age is not too srong a statement, because astronomical observations and theories have a direct bearing on man's view of his place in the universe. Thus the classical harmonies of the building at Greenwich mirror the ideas of Newton fully as much as the cenotaph which Boullee consciously designed as a memorial. In time the building acquired additional resonances with Newtonianism through the scientific work carried out there verifying the theory of universal gravitation. Associated with the massive masonry structures of Jai Singh is an Eastern astronomical tradition begun with the great pyramid of Egypt. Their very solidity mirrored a scientific tradition resistant to the scientific revolution sweeping over Europe from 1700 onwards. Mendelsohn's observatory symbolized the radical intellectual changes which began to flow from Einstein's ideas. Our own generation finds its view of the universe expressed by the shape of the solar observatory at Kitt Peak, a bold thrust into space.

The size, cost, and complexity of today's scientific instruments overshadow architectural considerations such as provision for ornamentation, light, and working space. These amenities have become subordinate to the housing's basic function of shielding the instruments from the elements. An observer using the 200-inch telescope on Mount Palomar must sit inside the telescope in an observer's "cage." Similarly, the shapes of the free-standing parabolic antennas of radio telescopes are more influenced by the laws of physics than the taste of architects. The resultant shapes are certainly imposing, but readers must answer for themselves the question, "Are they beautiful as well?"

Bibliography

Bennet, J.A., "Christopher Wren: Astronomy, Architecture, and the Mathematical Sciences," J. for the Hist. of Astronomy **6**, 149, 1975.

Blanpied, William A., "The Astronomical Program of Raja Sawai Jai Singh II and its Historical Context," Jap. Stud. History of Science **13**, 87, 1974

Cox, Warren, "The Observatories of Maharajah Sawai Jai Singh II," Perspecta, No. 6, 68, 1960.

Crowther, J.G., *Founders of British Science* (Cresset Press, London, 1940).

Donnelly, Marian C., *A Short History of Observatories* (Univ. of Oregon, Eugene, 1973).

Dorn, Harold and Mark, Robert, "The Architecture of Christopher Wren," Sci. Am., **245**, 160, 1981.

Giedion, Sigfried, *Space, Time and Architecture* (Harvard Univ. Press, Cambridge, 1967).

Howse, Derek, *Francis Place and the Early History of Greenwich Observatory* (Sci. Hist. Pub., New York, 1975).

Hutchings, Donald, Ed., *Late Seventeenth Century Scientists* (Pergamon Press, Oxford, 1969).

Kaye, George R., *The Astronomical Observatories of Jai Singh* (Varanasi Indological Book House, India, 1973) Reprint of 1918 edition.

Koenig. G.K., "Mendelsohn e l'Einsteinturm," Casabella , No. 303, 40, March 1966, No. 307,8, July 1966.

Laurie, P.S., "The Old Royal Observatory," Nat. Maritime Museum, H.M. Stationery Office, 1960.
Mendelsohn, Eric, *Three Lectures on Architecture* (Univ. of California Press, Berkeley and Los Angeles, 1944).
Rosenau, Helen, *Boullée and Visionary Architecture* (Academy Editions, London, 1974).
Summerson, J., *Sir Christopher Wren* (Archon Books, Hamden, Conn., 1965).
Talgeri, K.M., "Jai Singh and his Observatory at New Delhi," Sky and Telescope **18**, 70, 1958.
University of St. Thomas Dept. of Art, *Visionary Architects*, Catalogue (Univ. of St. Thomas, Houston, 1968).
Von Eckardt, W., *Eric Mendelsohn* (George Braziller, New York, 1960).
Whinney, Margaret, *Christopher Wren* (Praeger, New York, 1971).
Whitaker-Wilson, C., *Sir Christopher Wren* (Methuen, London, 1932).
Whittick, Arnold, *Eric Mendelsohn* (F.W. Dodge, New York, 1956).

Chapter 8

Scientists, Instrument Makers, and Artists in the Golden Age of Holland

WHEN THE DUTCH repudiated the Catholic rule of Spain in the seventeenth century, Holland emerged as a Protestant, mercantile republic and achieved unparalleled economic and political growth. At the same time instrument makers and scientists made significant advances in optics, mechanics, and electricity. The painters Rembrandt van Rijn and Jan Vermeer illuminated the world of art with their brilliance. How much were these artists influenced by the exciting scientific developments around them? Not only art museums but science museums contain clues indicating that the Golden Age of Dutch art resulted from a synergism which included the scientific revolution.

Early Optics and Art

The science of optics has many roots in seventeenth-century Holland. The telescope was invented there; a Dutch spectaclemaker, Hans Lippershey, applied for a patent on such a device in 1608. Willebrord Snell, a professor at Leiden, published his Law of Refraction in 1621. Among the more successful innovators in optics was Christiaan Huyghens. He was adept at lens grinding and polishing, and designed machines for these purposes. By using his improved lenses in telescopes of his own design he made a number of basic astronomical discoveries including Titan (a satellite of Saturn), the compact arrangement of four stars (the "Trapezium") in the inner part of the Orion nebula, and the true nature of Saturn's rings. Nor did the theoretical side of the subject escape his attention; the wave theory of light was one of his conceptions.

While Huyghens was making his basic discoveries in optics, Anton van Leeuwenhoek devised microscopes[7] which were the ultimate in simplicity. His instruments consist of two small plates of brass, silver, or gold, between which a spheroidal piece of glass

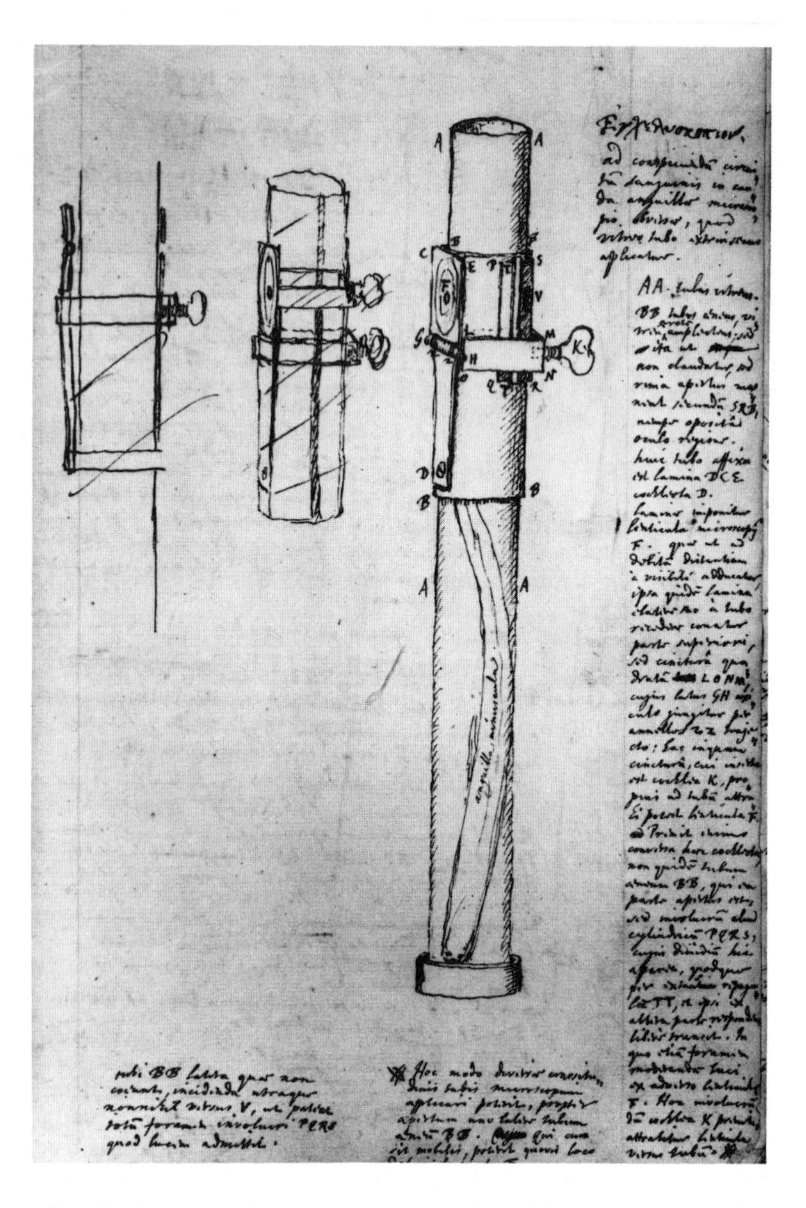

Fig. 43. Apparatus of Anton van Leeuwenhoek for observing the flow of blood in the capillaries of an eel's tail.

is clamped. An object to be viewed can be mounted on a pin manipulated by two screws.

One of Leeuwenhoek's favorite microscope subjects was the capillary system of a living fish's fin showing the transition of blood from arteries to veins, thus confirming Harvey's theory of the circulation of the blood. Among the numerous discoveries of Leeuwenhoek were the following: protozoa, striations in muscle fibers, bacteria in human spittle, yeast cells, structure in the lens of the eye, and sheaths of nerve fibers. No wonder scientists today consider him the founder of the science of microbiology.

What was the reaction of artists to these developments in optics? Certainly the invention of the telescope and microscope aroused considerable interest among the public, artists included. Artistic imagination was stimulated to the extent that several artists depicted optical instruments in their works, for example, Rembrandt's painting, *A Man with a Magnifying Glass*. In Antwerp in the first decades of the seventeenth century, a number of optical instruments appeared in works of art, the earliest being an engraving of a spectacle shop from *Nova Reperta* executed by Philipp Galle from designs by Johannes Stradanus. The great Flemish painter, Peter Paul Rubens, designed frontpieces for a number of books. An allegorical engraving by Thomas Galle from drawings by Rubens, depicting the principles of the photometer, binocular vision, and stereographic projection, prefaced each of the six books of the *Opticorum*, a treatise on optics by Francis Aguilon. Another example is the painting, *Sight*, by Jan Brueghel in which we see a telescope, magnifying glasses, spectacles, an astrolabe, and a number of other scientific instruments.

A curious link exists between the artist Jan Vermeer and the microscopist Leeuwenhoek. In 1676, Leeuwenhoek made his living as a clerk to the town bailiff of Delft, and in this capacity acted as receiver in the bankruptcy case of Catharina Bolnes, Jan Vermeer's widow. Did they meet while Vermeer was alive? We can only speculate, because there is nothing to show that Leeuwenhoek acted in this matter as anything more than a competent civil servant.

[7]The National Museum of the History of Science (*Rijksmuseum voor de Geschiedenis der Natuurwetenschappen*) in Leiden has several of Leeuwenhoek's simple microscopes, the most powerful magnifying 275 times and having a resolving power of 1.4 microns despite scratches on the lens. A model of one of Leeuwenhoek's microscopes is attached to a chain on the window sill in the microscope room so that a visitor may place this instrument close to the eye and try to view an insect wing or other object mounted behind the lens. The museum also possesses an objective lens ground by Huyghens and a number of clock models reconstructed from Huyghen's drawings.

Jan Vermeer and the *Camera Obscura*

So little is known about Vermeer from written records that he has been called the "Sphinx of Delft." The barest outline of his life is that he was married, had 11 children, lived in Delft, and besides painting worked as an art dealer, tavern keeper, and cloth seller. His total output numbers less than 40 paintings, a rarity which enhances his reputation.

If Vermeer were not affected by the tide of new discoveries in science, it would be surprising, yet evidence for scientific influence is scant. In his painting, *The Geographer,* one notices the expression on the subject's face; he appears to look beyond the confines of the simple room to the farthest reaches of the globe. Is he contemplating the intricacies of applying Gerhardus Mercator's projection to his map or merely thinking of shipping lanes to the Dutch East India Company's trading outposts? As in all of Vermeer's works the quality of the light is arresting; he depicts luminous, tranquil interiors which belie turbulent seventeenth-century Holland. The artist has taken an unusual point of view, barely above table height, on a line with the windowsill. In his *Officer and Laughing Girl* one has the same point of view. Although the officer is hardly a foot closer to the viewer than the girl, he is disproportionately larger, suggesting to many that Vermeer may have used a *camera obscura,* a device popularized by Giambattista della Porta in the sixteenth century. In its simplest form it is nothing more than a black box with a pinhole in one end, although an improved version would have a simple lens, inverting mirror and translucent screen. It would be a great convenience to an artist since it reproduced a three-dimensional scene in two dimensions, thus avoiding the use of complicated formulas for producing perspective drawings.

The Dutch art critic, P.T.A. Swillens, has painstakingly reconstructed Vermeer's studios from the artist's own accurately painted perspectives and known dimensions of floor tiles, chairs, and other depicted objects. The vanishing or viewing point in most of the analyzed paintings is about 8 inches higher than a table top—too low for the eye of a standing or sitting painter, but a natural height for a portable *camera obscura* placed on a convenient table. But no such device was listed in the inventory of Vermeer's worldly goods after his death, nor does the Rijksmuseum possess a contemporary example.

In some areas of his *View of Delft, Young Girl with a Flute,* and *Girl with a Red Hat* Vermeer used small dabs of paint or *pointillés* similar to the circles of confusion in an out-of-focus image in photography. Most likely the primitive lenses in early *camera obscuras* gave rise to this effect exploited so brilliantly by Vermeer. A sharper (i.e., increased depth of field) but darker image could be achieved by stopping down the lens.

Many seventeenth-century scientists wrote about the *camera obscura;* for example, Kepler in his *Ad Vitellionem Paralipomena* (1604) discussed sharpness of the image in this device relative to the size and shape of the aperture. The minutes of the Royal Society of London (1668) record Robert Hooke's demonstration of a *camera obscura* in front of that august body. The suggestive internal evidence from Vermeer's paintings plus the external evidence of widespread use of the *camera obscura* in the seventeenth century make it likely that Vermeer did use this aid.

Was Vermeer aware that under certain conditions the *camera obscura* can introduce distortions of natural perspective? Probably

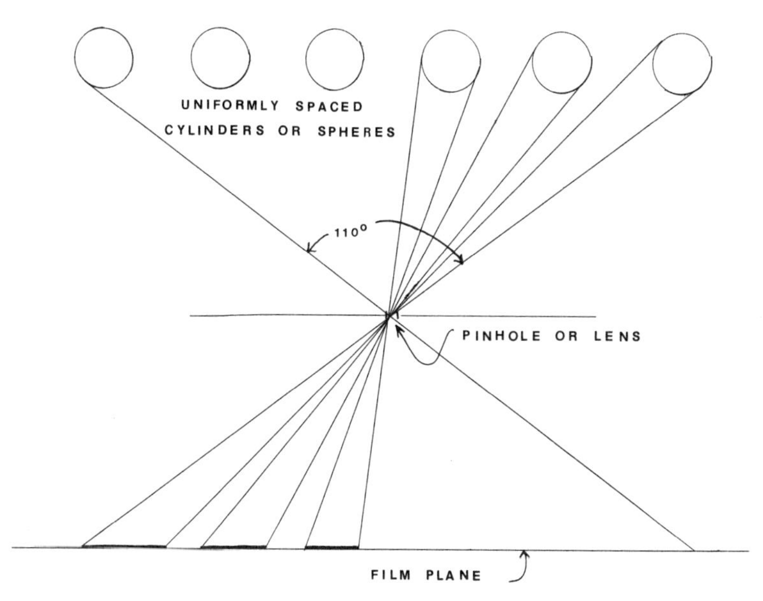

Fig. 44. Distortion introduced by camera obscura. *When the field of view exceeds Einstein's criterion (90 deg), destortions arise which are not easily compensated by the viewer. Note the increasing width of the images of the cylinders or spheres with increasing distance from the center.*

he was aware and avoided those extreme conditions, although in *Officer and Laughing Girl* he veers dangerously close to unnatural deformation. Einstein in a letter to M.H. Pirenne suggests that a spectator can compensate intuitively for deformation as long as an object subtends an angle of 90° or less. This rule of thumb corresponds to a photograph taken from a minimum distance always equal to half the height of an object. Amateur photographers often violate this rule by taking snapshots of people with outstretched limbs, resulting in extreme foreshortening. Swillen's reconstructions of Vermeer's rooms give viewing angles ranging from 29° to 65°, well short of Einstein's criterion. Another example of deformation, but one not arising in Vermeer's painting, occurs in a wide-angle view of a row of equally-spaced circular columns. The end columns appear fattened, as shown in the figure.

Today the photographic perspective is a commonplace, but in Vermeer's time the use of this device must have been a novelty, producing an effect which surprised and delighted his viewers. Thus science has subtly altered the way in which an artist perceives the world. The *camera obscura* was a starting point for artistic imagination but not the total explanation of Vermeer's art. In lesser hands the same scenes might have turned out to be lifeless and obviously the product of a copyist rather than a true artist.

Not only artists but poets were fascinated by the images produced by the *camera obscura*. About 40 years after Vermeer's death, an Englishman, John Gay, wrote in his poem, "The Fan,"

> Thus have I seen woods, hills, and dales appear,
> Flocks graze the plains, birds wing the silent air,
> In darken'd rooms, where light can only pass
> Through the small circle of a convex glass;
> On the white sheet the moving figures rise,
> The forest waves, clouds float along the skies.

Vermeer's *An Artist in His Studio* illustrates another aspect of the Golden Age. As in the previously discussed paintings, one notices the phenomenal character of the light from the unseen window and the perspective from a sitting rather than a standing position. The new elements are the brass chandelier and trumpet. The wealthy merchants of the country demanded beautiful furnishings for their homes, and well-made articles were supplied at a price by skilled craftsmen. The skills used in making a brass trumpet are similar to those needed for constructing a pneumatic pump, for instance. Thus the talents of these craftsmen became a basis for the advancement of science.

The van Musschenbroeks: Scientists and Instrument Makers

Just as a blend of theory and experiment is in large part responsible or the rise of western science, so a combination of scientist and instrument maker has also proven to be a fruitful collaboration. In extraordinary cases both of these talents may be possessed by one and the same individual, as for example by Newton. Einstein has written of Newton as follows:

Fortunate Newton, happy childhood of science! He who has time and tranquility can by reading this book live again the wonderful events which the great Newton experienced in his younger days. Nature was to him an open book, whose letters he could read without effort. The conceptions which he used to reduce the material of experience to order seemed to flow spontaneously from experience itself, from the beautiful experiments which he ranged in order like playthings and describes with an affectionate wealth of detail. In one person he combined the experimenter, the theorist, the mechanic, and, not least, the artist in exposition. He stands before us strong, certain, and alone: his joy in creation and his minute precision are evident in every work and in every figure. (from Foreword to *Opticks*, Dover 1952)

More often it takes the labors of two individuals to join theory and experiment. In the twentieth century, one thinks of Elmer A. Sperry's contribution of his high-intensity arc light and high-speed revolving mirror to the speed-of-light experiments of Albert A. Michelson, and of Stanley Livingston's inspired tinkering, which made E. O. Lawrence's cyclotron a success. Sperry was a highly-honored inventor by the time he was invited to design and construct apparatus for Michelson, but Livingston, then a graduate student, remained for years in relative obscurity while Lawrence won the laurels. Nobel prizes are awarded to scientists, not instrument makers. But the importance of skilled instrument makers should not be underestimated. For example, a solution to the problem of spherical aberration (failure of a spherical lens to bring all light incident upon it to a point focus) was worked out mathematically by Descartes as early as 1637, but it proved impossible to grind lenses to his thoretical hyperbolic curves in spite of various cleverly designed machines. As a result, scientists had to cope in other ways with the problem of lack of image definition from spherical lenses. Mechanical design of telescopes was equally important, for an observer could not use an excellent lens to full advantage if it were improperly centered, or if the mechanical arrangement for focusing were imprecise.

In the seventeenth century the interaction between instrument makers and their customers took place not only in the instrument makers'shops, but also in coffee houses and private homes. Hardly a day passed when Robert Hooke's diary did not record a visit to an instrument maker. For example, the record of December 12, 1675, includes a discussion with the celebrated clockmaker Thomas Tompion.

Discoursed with Tompion. Borrow Clock. About bellows new invented and about oval watch. Described the flap of his barometer bolt head.

Before the founding of scientific societies such contacts between instrument makers and their customers were important focal points for the informal exchange of scientific information

In seventeenth-century Holland a family of instrument makers and scientists unique in the history of science arose. The founder of this dynasty, Joost van Musschenbroek (b. 1614), was a brass founder and maker of church chandeliers. His two sons, Samuel and Johan, became scientific instrument makers. Samuel was 21 years older than Johan and taught him the craft. Certainly Samuel, and perhaps also Johan, worked for Huyghens. They must have been competent lens grinders and probably ground and polished the lenses of their own microscopes. In the National Museum of the History of Science at Leiden is a single-barrel air pump (1675), the only extant instrument of Samuel's and a splendid example of its type. The mahogany table supporting the glass bell jar has Italianate legs, while the brass pump has legs and handles typical of the best domestic furnishings of the time. It bears Samuel's (and later Johan's) workshop mark: two crossed keys and an oriental lamp.

Because air would obviously leak if there were ill-fitting parts, the construction of an air pump for laboratory use imposed a new

Fig. 45. Ornate single-barrel air pump made by Samuel van Musschenbroek (1675).

Table 11.

van Musschenbroek Family Tree

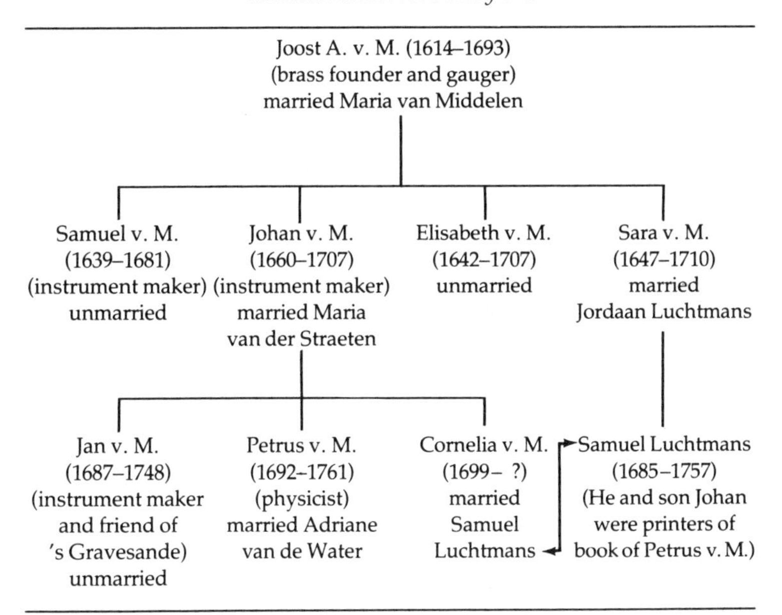

order of precision upon instrument makers. Some earlier instruments had been of enormous size (the quadrant and telescope, for example) with a few moving parts, but air pumps were both large and complex. Each of the succeeding van Musschenbroek instrument makers improved the design, and in the hands of Jan van Musschenbroek it became an item of commercial production. Some of the techniques devised by the air-pump pioneers (besides the van Musschenbroeks there were Boyle, Huyghens, Papin, Hauksbee, and others) were later used in the development of the steam engine.

Johan had two sons: Jan and Petrus van Musschenbroek. The younger son, Jan, was trained in the art by his father, and became the last of the instrument makers of this remarkable family. The family register states that he read Latin, Greek, English, French and German, and that he was also a "learned and wide-awake mathematician." His career is closely linked with that of the University of Leiden mathematics and astronomy professor, Wilhelmus Jocobus 's Gravesande (1688-1742). 's Gravesande had met Newton and had been made a Fellow of the Royal Society of London, later winning fame as an outstanding educator. Following the example of Galileo, he believed that a course in experimental physics should be accompanied by demonstrations of the physical principles involved. To that end he had Jan van Musschenbroek build a series of instruments[8] for the demonstration of the phenomena of the entire physics of his time. The instruments were 's Gravesande's personal property as was the custom, but were bought by the university at his death. Since he did not sign his instruments and for the most part did not advertise, Jan probably was a modest man, allowing his excellent craftsmanship to speak for itself. His fame was spread by the publication of a physics textbook by 's Gravesande (1720) which, by the third edition (1742), had expanded to two volumes and included 127 fine copper engravings of the van Musschenbroek instruments. The more complicated ones having the look of imposing pieces of furniture are shown in considerable detail, apparently to facilitate copying. Thus began a tradition of physics teaching by means of lecture-demonstrations later exploited skillfully by Faraday in his famous lectures at the Royal Institution in London.

Jan's younger brother, Petrus van Musschenbroek, became interested in physics while studying under 's Gravesande and eventually became professor of physics at Leiden in 1740. He too emphasized the experimental method, and to him two inventions are credited: the pyrometer and the Leiden jar. His pyrometer was a device to measure the relative expansions of different metal rods all held at the same temperature, the dilations being amplified by a system of cogwheels and registered on a round dial. Indeed, a very similar method is used to instruct many beginning physics students today. Petrus was preceded by a few months in his invention of the Leiden jar by von Kleist, a cathedral dean at Kamin, Germany. But von Kleist never told why he performed his experiment, and never sufficiently publicized it. Petrus' experimental apparatus could hardly be simpler: an ordinary glass jar filled with water and a stopper which admitted a wire from an electrostatic charge generator. The idea behind the experiment was that a conductor in an enclosed, nonconducting vessel might lose its charge less rapidly than a conductor in open air. An assistant supplied by accident the missing ingredient, an outer conducting layer, when he held the charged vessel in one hand, and touched the wire with his other hand, receiving a fearful shock. Later, a metallic foil wrapping became the outer conductive layer. Petrus repeated the experiment himself and is reported to have written that he "would not take another such shock for the whole kingdom of France." Surprisingly, such a

[8]One can find duplicates of many of these instruments at the Teyler's Museum in Haarlem and the Science Museum in London.

simple apparatus could store electricity and deal powerful blows. Soon everyone in European scientific circles was trying to repeat the *"expérience de Leyde,"* as Abbe Nollet called it.

The 1740's represent the heyday of electrostatic experimentation. Park Benjamin in his *The Intellectual Rise in Electricity* (1895) describes it as follows:

> . . . Both Hansen and Bose found, at about the same time, that not only could a practically continuous supply of electricity be obtained [with a glass sphere rotated by a crank], but one of much greater strength than had hitherto been known. Hansen suspended a boy with his toes in proximity to the globe, and drew sparks from his fingers. Bose disposed twenty soldiers in line, with hands touching and administered a shock to all of them at once . . . No one knew better the art of playing to the gallery . . . He sets jets of fire streaming from electrified objects and exhibited them to people who flocked to his laboratory. He invited guests to an elegant supper-table loaded with silver and glass and flowers and viands of every description, and as they were about to regale themselves, caused them to stand transfixed with wonder at the sight of flames breaking forth from the dishes and the food and every object on the board. The table was insulated on pitch cakes, and received the discharge from the huge glass retort which revolved in another room. He introduced his ardent pupils to a young woman of transcendent attractions and as they advanced to press her fair hand, a spark shot from it accompanied by a shock which made them reel. Others, who had the boldness to accept his challenge to imprint a chaste salute upon the damsel's lips, received therefrom a discharge which Bose says "broke their teeth," but Bose here either exaggerates more than usual, or else neglects to explain how the young lady bore her share of the injury.

The popular imagination had not been so captivated by a scientific toy since the invention of the microscope and telescope, but modern sensibilities may be offended by the description of uncontrolled experimentation with human subjects, however willing. The tooth-breaking incident should probably not be accepted literally for the reason Park states, but a tooth's nerve could nevertheless be given a powerful and uncomfortable electrical stumulus in this manner. To test whether a nerve is dead or not, dentists use the same method today on a reduced scale.

The design of scientific instruments passed through several stages, beginning with rudimentary, then highly ornamented, and finally utilitarian shapes. This evolution parallels the changing styles of observatory architecture examined in Chapter 7, but instrument design evolved faster because designers had a more flexible medium with which to work, and instruments had to be designed before the buildings housing them. Ornamentation reached its zenith in the elaborate silver microscope made (1761) by the gifted English instrument maker, George Adams, for King

Fig. 46. An eighteenth-century electrostatics demonstration (from Abbe Nollet's "Ricerche sopre le cause particolari di fenomeni elettrici.")

George III. One may see this magnificent work of art (can it actually have been used as a microscope?) encrusted with urns, fluted columns, and statuettes, at the museum of the History of Science in Oxford. As royal patronage gradually declined, designers realized that instruments did not have to imitate the shapes of furniture or Greek temples in order to be beautiful; they began to allow the form of the instrument to reveal its function.

Fig. 47. An elaborate silver microscope made by George Adams (1761) for King George III.

With the advent of the Leiden jar, scientific instruments shed the last vestige of ornament. The shapes are in general utilitarian, even ugly, although Utrecht University Museum contains an exceptional one covered outside by red lacquer with a representation of Chinese figures fishing and walking among temples.

Leiden jars were often attached to electrostatic generators to make their discharge more powerful. These ungainly bottles stand in a direct line of evolution leading to the electrostatic accelerators (Van de Graaffs) and cyclotrons used today in nuclear research and medical treatment. Modern forms of Leiden jars, known as capacitors or condensers, are found in virtually every electronic apparatus.

In the nineteenth century, James Clark Maxwell (1831–1879) synthesized the work of the early electrical and optical experimenters in the following four equations:

$$\nabla \cdot \mathbf{E} = 0$$
$$\nabla \cdot \mathbf{H} = 0$$
$$\nabla \times \mathbf{E} = -\mu \dot{\mathbf{H}}$$
$$\nabla \times \mathbf{H} = \epsilon \dot{\mathbf{E}}$$

The first two equations express Carl Friedrich Gauss' laws of electricity and magnetism, respectively. The third equation sums up Michael Faraday's law of induction, which gives the electrical effect of a changing magnetic field. The last equation embodies Andre Marie Ampēre's law (later extended by Maxwell), which gives the magnetic effect of a changing electric field. Both the language used above in describing the equations and their typographical appearance reveal certain symmetries, reflecting the roles played by electricity and magnetism in nature. Because of this symmetry, plus their conciseness and scope, these equations rightly deserve to be called elegant. They constitute the fundamental principles of electricity, magnetism, and light basic to the operation of Leiden jars as well as all the modern electrical instruments including oscilloscopes, accelerators, lasers, etc. Unlike the handsome van Musschenbroek apparatus, the equations stand austere and abstract with a beauty not readily appreciated by the uninitiated, a situation in which modern abstract art also finds itself. Some ways in which science has directly influenced art will be discussed in the next chapter.

Bibliography

Bell, A. E., *Christian Huygens and the Development of Science in the Seventeenth Century* (E. Arnold & Co. London, 1947).

Bernal, John D., *The Extension of Man* (The M.I.T. Press, Cambridge, 1972).

Chew, V.K., *Physics for Princes: The George III Collection* (Her Majesty's Stationery Office, London, 1968).

Crommelin, C.A., *Descriptive Catalogue of the Physical Instruments of the 18th Century* (National Museum of the History of Science , Leiden, 1951).

____, "Physics and the Art of Instrumentmaking at Leyden in the 17th and 18th Centuries," in *Lectures on Physics and Physiology* (A.W. Sijthoff, Leyden, 1926).

Daumas, Maurice, *Scientific Instruments of the Seventeenth and Eighteenth Centuries,* translated and edited by Holbrook, Mary (Praeger, New York, 1972).

Koningsberger, Hans, *The World of Vermeer 1632-1675* (Time Inc., New York, 1967).

Lavên, W. J. and Cittert-Eymers, J. G. van, *Electrostatic Instruments in the Utrecht University Museum,*Descriptive catalog (Utrecht, 1967).

Michel, Henri, *Scientific Instruments in Art and History* (The Viking Press, New York, 1967) Translated by R. E. W. Maddison and Francis R. Maddison.

Pirenne, M.H., *Optics, Painting and Photography* (Cambridge Univ. Press, 1970).

Rooseboom, Maria, *Microscopium* (National Museum of the History of Science, Leiden, 1956).

____, *Bijdrage tot de Geschiedenis der Instrumentmakerskunst in de Noordelijke Nederlanden tot Omstreeks 1840* (Rijksmuseum voor de Geschiedenis der Natuurwetenschappen, Leiden, 1950).

Schwarz, Heinrich, "Vermeer and the Camera Obscura," Pantheon, **24**, #3, May-June 1966.

Seymour, Charles, Jr., "Dark Chamber and Light-Filled Room: Vermeer and the Camera Obscura, " Art Bulletin **46**, 323, Sept. 1964.

Singer, Charles, Holmyard, E.J.,Hall, A.R., and Williams, Trevor I., *A History of Technology,* Vol. III (Oxford, at the Clarendon Press, 1957).

Swillens, P.T.A., *Johannes Vermeer* (Spectrum Publishers, Utrecht, 1950).

Chapter 9

Science and Art

THIS CHAPTER CONCERNS an artist who faithfully rendered scientific apparatus (Wright), three modern surrealists whose works have intriguing parallels with modern physics (Escher, Magritte, and Dali), the inventor of high-speed photography and father of the motion picture (Muybridge), and a scientist who is also an artist (da Vinci)—a small sample from the world of art illustrating several direct interactions between art and science.

Joseph Wright of Derby

Except for a trip to Italy during which he painted volcanic eruptions, satyrs, and grottoes, Joseph Wright lived in Derby, the town in which he was born. The bulk of his work consists of routine portraits and landscapes commissioned by well-to-do families of Derbyshire. For other works his favorite device is the scene illuminated by candlelight, an artifice also prominent in the work of the French painter Georges de la Tour. Wright's sources of light are usually obscured by a letter, balloon, or other object rendered translucent in the painting, but occasionally the light source is visible, such as the red-hot iron bar in *The Blacksmith's Shop* (1771). In less skillful hands this device would have become hackneyed, but Wright manages to introduce variations that hold one's interest. Three paintings of this genre show the use of scientific instruments. The first two paintings depict Wright's contemporaries viewing demonstrations used in the popularization of science, while the last is the romantic re-creation of an earlier discovery.

A Philosopher Giving a Lecture on the Orrery (c. 1764-1766). The first orrery, the eighteenth-century equivalent of the planetarium, was constructed in 1709 by the clockmakers George Graham and Thomas Tompion. The name is taken from Charles Boyle, the fourth Earl of Orrery, who commissioned John Rowley to make one for him in about 1712. In Wright's picture, the elder boy's sleeve obscures the source of light representing the sun. Mercury and Venus are too close to the sun to be visible in the picture, but one sees all the other planets from Mars on the left through Saturn on the right. The moon tracks around the earth on a circular frame. The planet Jupiter with its four revolving moons draws the attention of the boy facing the viewer. Not shown is a handle or other mechanism to make the orrery turn and reproduce all the motions of the heavenly bodies with their correct relative speeds. On the inner rim of the horizontal circle representing the plane of the ecliptic the signs of the zodiac are inscribed; the word "Cancer" is visible in the painting. The months of the year are on the outer rim; one can see "July" written there. Someone seems to have left a glass of wine on this circle. The inclined circle is the Equator, and the great circles at right angles to it terminate in the celestial pole. Taken together these hoops form an armillary sphere. With this mechanism the natural philosopher standing behind the orrery can demonstrate eclipses, the phases of the moon, the phases of Venus, the changing seasons, etc. The artist overlooked one infelicity in the composition: the arm of the man taking ntoes appears to rest on the broad brim of the lady's hat.

An Experiment on a Bird in the Air Pump (c. 1767-1768). The vacuum pump developed by the van Musschenbroeks was dis-

Fig. 48. A Philosopher Giving a Lecture on the Orrery *by Joseph Wright. Derby Art Museum.*

cussed in the previous chapter. Wright's painting shows another type of vacuum pump invented by Francis Hauksbee about 1709. The handle drives a rack and pinion mechanism for moving the pistons in the cylinders. In the painting the rack on the right-hand cylinder is raised, indicating a vacuum on that side. A valve is then turned by hand and the piston on the left is raised. The pistons are moved in succession, eventually exhausting the air from the glass receiver through a copper tube. The demonstration being performed is not for the squeamish. As the air is exhausted, a dove in the receiver at first collapses, then is convulsed, and finally dies. One supposes that the demonstrator in the painting will not allow the experiment to proceed so far, because his hand is poised upon the valve at the top of the receiver. When his assistant holding a watch in his hand gives the signal, he will open the valve admitting air into the chamber, thus reviving the bird. The boy at the right is lowering the bird's cage so that the bird can be replaced in its domicile. On the table are a pair of spectacles in a case, a candle snuffer, a cork, a large glass containing an unknown liquid, and a pair of Magdeburg hemispheres for further experiments. These hemispheres take their name from a celebrated experiment which Otto von Guericke performed for Emperor Ferdinand III at Regensburg in 1654. Two large copper hemispheres about 22 inches in diameter were placed together forming an air-tight joint. When the air was pumped out, a team of eight horses pulling on each side of the sphere could not pull it apart. The force due to the atmosphere on such a large sphere amounts to nearly three tons, but the smaller ones shown on the table in Wright's picture could be pulled apart more easily.

The Alchemist in Search of the Philosopher's Stone Discovers Phosphorus (1771). During the alchemists' search for the Philosopher's Stone which could turn base metals to gold, and for the Elixir of Life which could cure all ills and prolong life indefinitely, real scientific discoveries were occasionally made, such as the accidental discovery of phosphorus by Hennig Brandt in Hamburg about a hundred years before this painting was made. While heating a residue of evaporated urine he found a substance which glowed in the dark and caught fire spontaneously when exposed to air. The details of the process were kept secret until the early part of the eighteenth century. This painting depicts the moment of discovery. It shows a brick furnace on the right into which the materials for making phosphorus have been placed. After strong heating, the phosphorus is distilled into the glass retort shown one-third filled with water. The phosphorus vapor glows with a bluish light and is ejected in a luminous jet from a hole in the top of the retort, causing the alchemist to kneel in awe.

Fig. 49. The Alchemist in Search of the Philosopher's Stone Discovers Phosphorus *by Joseph Wright. Derby Art Museum.*

A youth in the background is using a blowpipe on a bed of charcoal, but is distracted from his task by the alchemist's discovery. Throughout the picture Wright depicts antique apparatus with great care.

M.C. Escher and the Reflection of Light

In the last few years, the graphic work of the Dutch artist Maurits C. Escher has become popular with the general public, but it has always held a special fascination for scientists. Mathematicians appreciate his tessellations and Möbius strips which pose topological problems in artistic form. Indeed, Escher has said that he often feels closer to mathematicians than to his fellow-artists. Psychologists may find in some of his prints illustrations of common optical illusions, while solid state physicists may come across striking examples of symmetry, group theory, and crystallographic laws.

Escher made a series of seven prints dealing directly with the properties of reflected light and revealing an unsurpassed technical virtuosity in print-making. In the lithograph *Three Spheres II* (1946) one sees a crystal, a silvered, and a translucent ball resting on a flat, smooth surface. Escher perfectly renders the quality of light from each one, showing the mutual reflections of the spheres in each other as well as the reflection of the artist making the print. One sees the artist and his room more clearly in *Hand with Reflecting Globe* (1935), but the room and its simple furnish-

Fig. 50. "Three Spheres," lithograph by M.C. Escher

ings appear grossly distorted, imparting a nightmarish quality. The tranquility that one finds in a Vermeer interior is gone, and a vague feeling of unease has intruded itself. The dark undercurrents of life may be symbolized in the enormous fish swimming just under the surface in *Three Worlds* (1955). The reflections of gaunt tree limbs and autumn leaves floating in the water com-

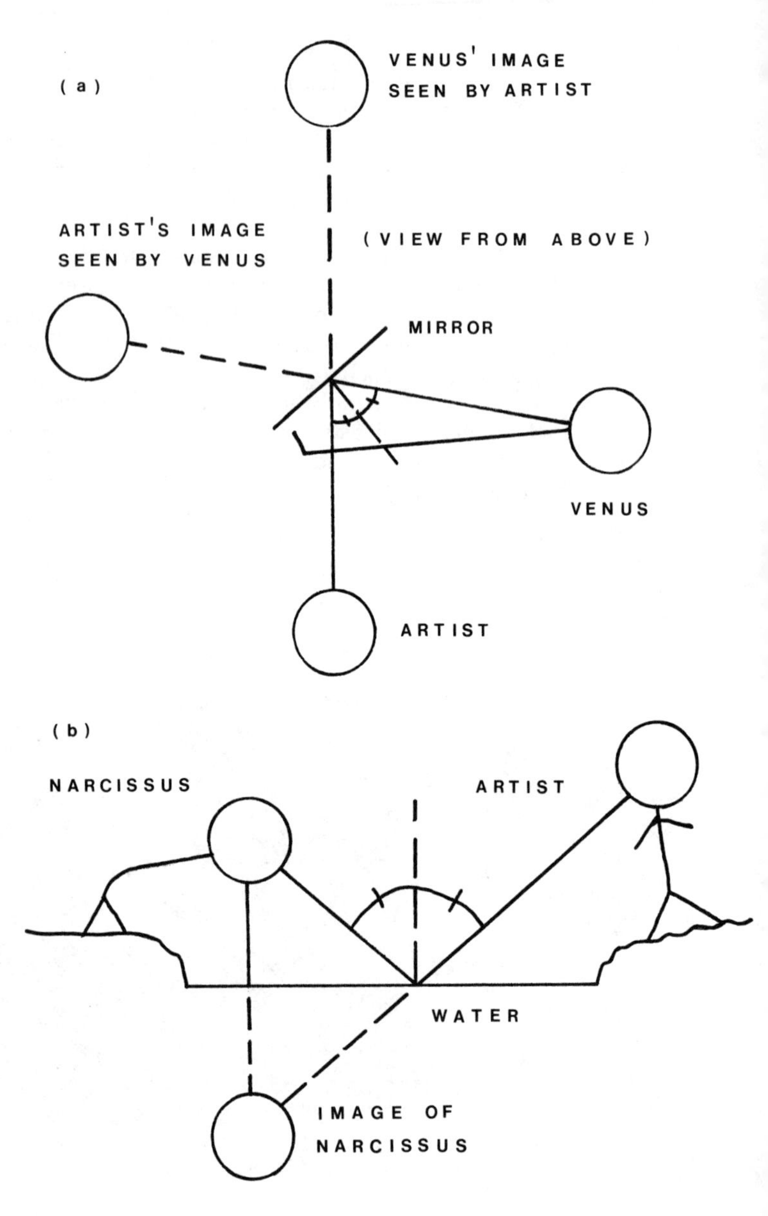

Fig. 51. (a) If the artist sees Venus in the mirror in Velazquez' painting, Venus and Cupid, *then Venus must see the artist according to the laws of optics, but contrary to the usual use of cosmetic mirrors.*

(b) In this case, both artist and Narcissus can see Narcissus' image in the water.

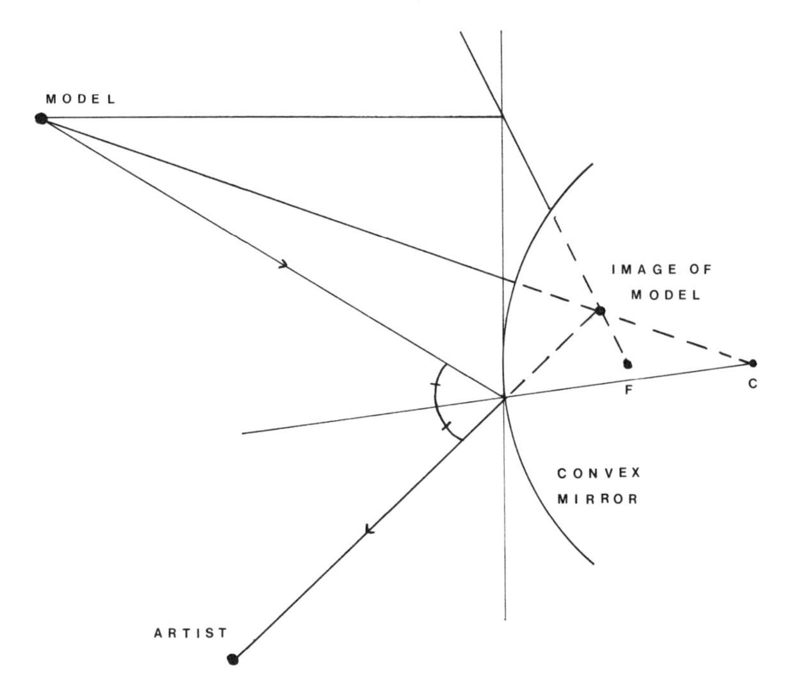

Fig. 52. Both artist and model can see the model's image in a convex mirror. "F" is the focal point of the mirror, and "C" is its center of curvature.

plete a picture which conveys an atmosphere of despair and weariness.

Escher's studies of light incident upon various types of globes follow the tradition of several earlier artists. Quentin Matsys' painting *The Moneylender and his Wife* (early sixteenth century) contains a small convex mirror which upon close inspection reveals a reflection of the artist at work. On the wall behind the couple in Jan Van Eyck's painting *Giovanni Arnolfini and his Wife* (c. 1434) is a handsome convex mirror. Van Eyck's detailed pictures necessitated an examination of the properties of light; for example, his painting of the *Annunciation* (c. 1425) depicts the angel's wing in the sequence of the colors of the rainbow. The light from the stained-glass windows at the upper left is indicated in the form of tangible rays directed toward the Virgin. In Egyptian art of the Amarna period, court artists drew similar rays from the sun terminating in ministering hands; such was the beneficence the sun shed on the Sun-God, Akhnaten. Escher's fellow-countryman, Vermeer, painted an unsilvered crystal globe (a household ornament) suspended on a ribbon as the visual focus

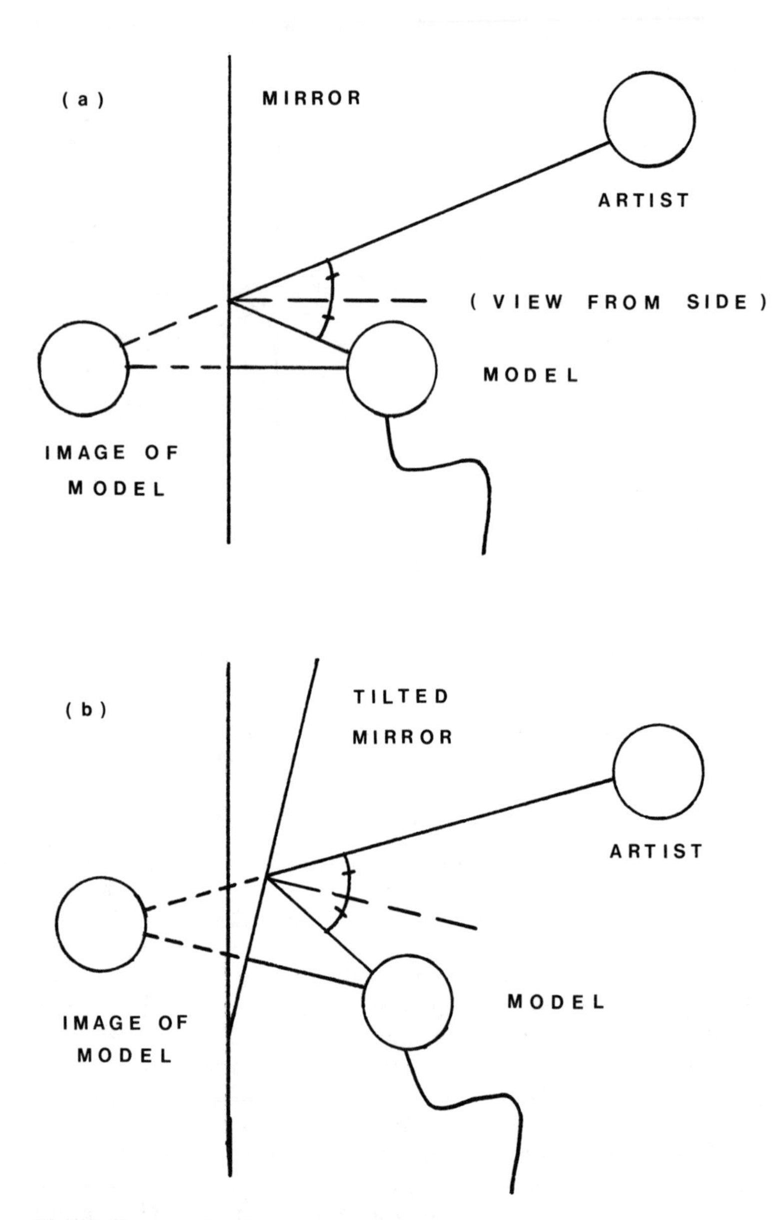

Fig. 53. Ingres painted Mme. De Sennones *from above, so her mirror image must appear* above *her head whether the mirror is vertical (a) or tilted (b). Ingres incorrectly paints the image* below *her head.*

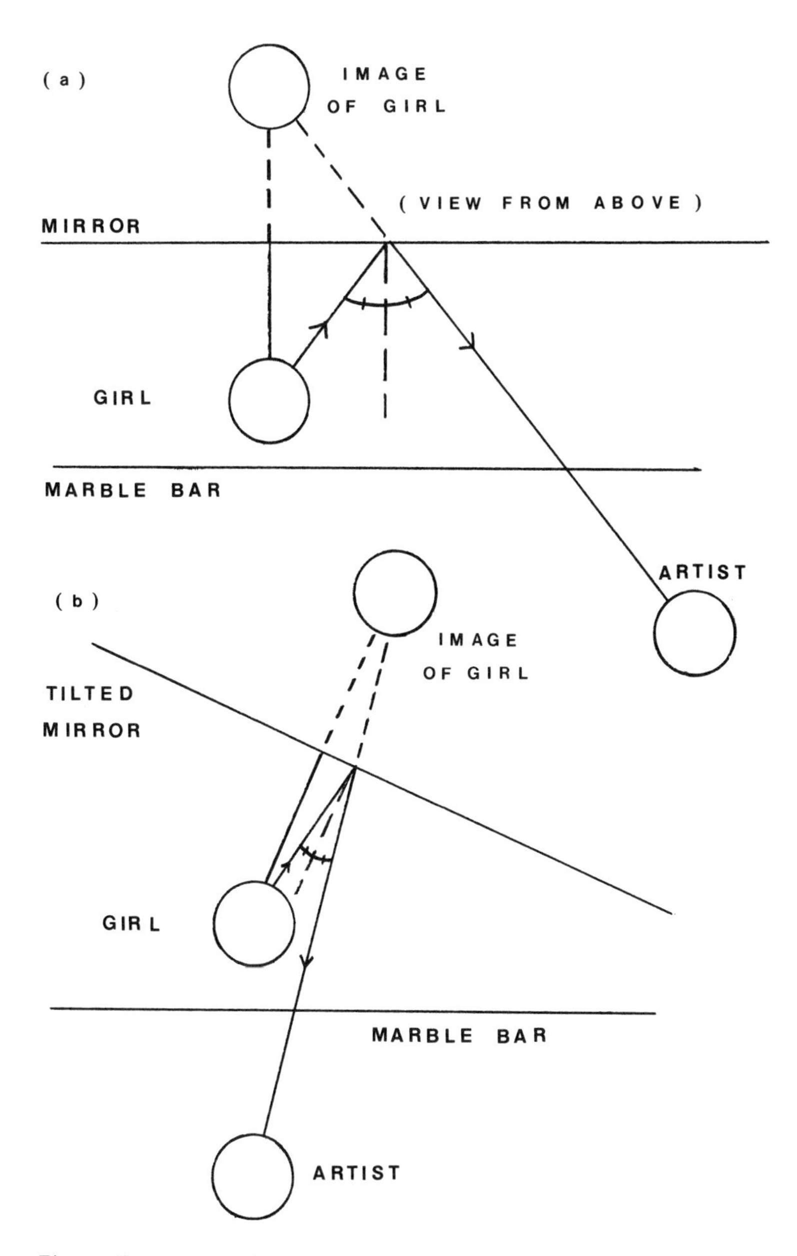

Fig. 54. Two ways in which the image of a girl in Manet's Bar at Folies Bergêres *could appear on her right in a background mirror. But the top view contradicts the artist's frontal position, while the bottom has the mirror at an angle contrary to the bar's parallel reflection. Manet evidently disobeyed the laws of optics for his own artistic purposes.*

of the upper right hand corner of his *Allegory of the New Testament*. Somewhere in its reflection must be hidden the image of the elusive Vermeer.

Not every painter has been as successful as Escher, van Eyck, Matsys, and Vermeer in representing reflections, or rather, other artists have freely distorted the laws of geometrical optics to suit their artistic purposes. Consider the painting *Venus and Cupid* by Diego Rodriguez de Silva y Velasquez. Cupid appears to hold a flat mirror so that Venus may view herself, but actually Venus must see not herself but the artist, since the artist sees her reflection. A ray diagram based on the Law of Reflection (the angle of incidence must equal the angle of reflection) makes this clear. This error (I will call it "reciprocity failure") is so common throughout the history of art that it has become an artistic convention accepted by artist and viewer alike. Reciprocity failure is all the more remarkable because it occurs in pictures which are otherwise extremely realistic. The table lists a number of works of art exhibiting reciprocity failure, but any museum should have at least one example. Pictures entitled *Toilette of Venus* or *Bathsheba* are good candidates for scrutiny, but *Narcissus* is not, because the

Table 12.

Examples of Reciprocity Failure

Artist	*Title of Work*	*Location*
Carracci	*Perseus Slaying the Medusa*	Farnese Palace, Rome
Janssens	*Allegorie des Gesichtssinnes*	
Raffael	*Allegorie der Prudentia*	Vatican, Rome
Memling	*Die Eitelkeit*	Strassburg
Titian	*Die Spiegelseherin*	Vienna
Goya	*Hasta la Muerta*	
Einsle	*Junge Frau*	Germany, private collection
Lütgendorf-Leinburg	*Dame vor dem Spiegel*	Würzburg
Couture	*Vor dem Spiegel*	Munich
von Mieris	*Dame vor dem Spiegel*	Munich
Terborch	*Dame an Putztisch*	Paris
Daumier	*Der Schauspieler*	
Maratta	*Bathsheba*	Vienna
Rembrandt	*Bathsheba*	Leningrad
Rubens		Vienna
Cagliari		Basil
Tintoretto	*Toilette of Venus*	Munich
Titian		Washington
Valesquez		London

usual reflecting pool is extensive enough to allow both artist and Narcissus to see Narcissus' reflection (see diagram). Similarly pictures showing convex mirrors are not usually examples of reciprocity failure, as the ray diagram shows.

Another type of error in depicting reflections is the shifting of the image position in the painting from where it would appear according to the laws of optics. I will mention three examples of position-shift error. In Jean August Dominique Ingres' portrait of *Mme. De Senonnes* the image of the model in the mirror appears *below* her head. The ray diagrams show that for either a vertical or tilted mirror an artist viewing the model from above must see an image *above* the model's head. If Ingres had strictly followed the laws of optics, the image would be higher than the model, dominating the portrait and spoiling the composition. Another of Ingres' portraits, *Mme. Moifessier,* correctly shows a mirror reflection lower than the model because in this instance the artist's viewpoint is from below.

A second example of position shifting occurs in Edouard Manet's *A Bar at the Folies-Bergères* where the reflection of the serving girl is pictured on the right side of the painting. This could occur in two ways as shown in the ray diagrams—either the artist was standing on the right side or the mirror was tilted about a vertical axis. But the artist gives us a frontal not a side view of the serving girl, and the reflection of the bar's marble top is parallel to itself, ruling out these two explanations. Since the reflection shown is a physical impossibility, why did the artist distort in this way? A clue is given by the reflection of the man in evening dress with whom the girl must be talking. Evidently Manet's intent was to give the impression that the girl is talking to the *viewer,* that is, the viewer is standing where the man in evening dress must be.

The last example of position shifting (in this case a rotation) is the window mullion reflection seen in a wine glass in William Hogarth's engraving *A Modern Midnight Conversation.* The mullions should be vertical in the reflection since they are presumably vertical in the window. This error seems to serve no artistic purpose; it probably represents an observational lapse.

René Magritte and the Substantiality of Matter

The enigmatic visions in René Magritte's paintings have the power to shock and amuse us. In many ways Magritte is performing the same visual sleights of hand in paint that Escher did in other media.

At first glance his *Signature in Blank* (1965) appears to be a scene of an elegant horsewoman riding through a woodland, but one soon realizes that the horsewoman is in a different plane from her mount. Magritte has conjured up a universe in which trees may pass through seemingly solid objects, rather like the smaller-scale world of quantum mechanics where seemingly solid particles such as electrons may pass through apparently insurmountable barriers. Since in this theory matter can be described by waves, it follows that these waves may sometimes be able to penetrate obstacles as easily as light waves penetrate glass. Magritte did not state such scientific references explicitly, but to a large degree he seems to have unconsciously absorbed the viewpoint of twentieth-century physics and expressed it in compelling visual symbols.

In *The Postcard* one has the unexpected juxtaposition of an enormous green apple looming over a man's head. The apple, mysteriously floating in mid-air, appears more massive than the mountain scenery to which the man devotes his attention. For the viewer this gravity-defying act may bring associations with Newton's theory of gravitation, the inspiration for which is said to have been an apple falling on Newton's head.

Magritte deliberately challenges our conventional way of thinking about the substantiality of matter. One might suppose that most painters unconsciously absorb and incorporate into their art the contemporary prevailing theory of the universe and its implications concerning the nature of matter and ponderability. To demonstrate this hypothesis, I have chosen several examples of paintings of the Madonna, a genre in which a painter inevitably must convey his idea of a human being's (that is, the Virgin Mary's) place in the universe.

The Madonnas in medieval paintings often have a kind of floating quality, for example *Madonna Enthroned* (thirteenth century) by Cimabue and the picture of the same title by Duccio. In contrast to Magritte's massive apple, these figures lack weight, since the characterization is entirely two-dimensional, contributing to a feeling of other-worldliness. The floating Madonna is in harmony with the Ptolemaic picture of the earth as center of the universe. According to Aristotle the "natural" motion of earthly bodies was toward the center of the earth; air and fire on the other hand possessed the quality of lightness and were drawn upward. In the Christianized version angels and other spiritual beings rose toward the Empyrean as their "natural" sphere of being, but the introduction of Copernicanism removed the props from this type of thinking. Copernicus thought gravity was a property of all spherical bodies, the most perfect form in nature, and "naturally" other bodies sought to unite themselves to this perfect form. According to this view, a falling leaf sought to join itself to the spherical earth.

In contrast to earlier pictures, Raphael's images of the

Madonna give every impression of weight and solidity. Consider his *Madonna in the Meadow* (1505) where the figure of the Virgin is firmly earth-bound: she is seated on the earth, she glances downward at her child, and flowers spring up around her. Of course many influences other than Copernicanism are at work here; technical progress in representing three-dimensional bodies on two-dimensional canvas and a religious climate which tolerated an emphasis on the human qualities of Mary favored this kind of tangibility. Yet it seems more than coincidental that Raphael painted during the lifetime of Copernicus when the old ideas of gravity, weight, and density were being re-examined.

The Tychonian scheme of the universe which again placed the earth at the central position was in that sense a revival of medieval beliefs. At about the same time El Greco used certain medievalisms in some of his works; for example, his painting *The Immaculate Conception* (1607–1613) recalls late Byzantine art in its suppression of space. The elongation of the figures suggests an upward movement toward heaven, and a circle of angelic heads stands for a large crowd. Again a rather striking coincidence exists between developments in science and art.

Salvadore Dali and Atomism

The titles of some of Salvadore Dali's paintings reveal a conscious effort to incorporate the terms and ideas of modern physics into his work. Consider, for example, the following: *The Maximum Speed of Raphael's Madonna, Anti-Protonic Assumption, Saint and Three Pi-Mesons, Intra-Atomic Equilbrium of a Swan's Feather, Madonna in Particles,* and *Nuclear Cross.* How much of this is posturing and obfuscation, and how much represents real insight? One suspects that the artist may lack a deep understanding of an anti-proton or a pi-meson, but at least one fundamental idea has taken hold of artistic imagination—the concept of *Atomism.* He writes in *Dali on Modern Art,*

> The most transcendent discovery of our epoch is that of nuclear physics regarding the constitution of matter. Matter is discontinuous and any valid venture in modern painting can and must proceed only from a single idea, as concrete as it is significant: *the discontinuity of matter.*

The opposite of Atomism is the concept of the *Continuum.* Because energy can be transported in only one of two ways, either by particles (Atomism) or by waves (Continuum), most of the developments in science can be classified in either the Atomism or Continuum category. But some of the most significant scientific advances have creatively combined particle and continuum concepts, as in quantum mechanics, for example. Although the human mind evidently finds the particle and continuum pictures congenial, reality is undoubtedly more complex than either picture. Dali has given us a clue that these two concepts from physics may furnish a useful vocabulary for art criticism.

Most art possesses approximately equal particle-like and continuum-like properties. For example, a sculpture is an object-in-itself like an atom, but it usually has a continuous surface. Some art can be classified as predominantly particle-like or continuum-like. Dali's *Madonna in Particles,* showing the body of the Virgin disintegrating flight, is predominantly particle-like, but the ultimate spiritual unity to be achieved in Heaven is also strongly implied. Similarly, the pointillism of Georges Seurat and the cubism of Georges Braque are based upon particle-like techniques, but their paintings can easily be viewed as a unity. The ultimate particle picture is probably Escher's *Concave and Convex* which is divided along a vertical line through the middle into two different perspectives. On the left we look *down* at the scene, while on the right we look *up.* Since one human brain can never reconcile the two different ways of perceiving the same scene at the same time, the picture has two irreducible elements.

Picasso's particle-like picture *Saltimbanques* (1905) has no single, consistent viewpoint, because the children are viewed from above, while the seated woman is seen at about eye level. In thus moving through space the artist has also moved through time, or rather the artist has taken several views made at different times and combined them into one integrated whole. In this way his painting is distinguished from a photograph which normally records an instant of time. Strangely enough, this seemingly modern notion of space-time has been anticipated in many medieval paintings. To the medieval mind, Christ with his saints and angels dwelt in eternity rather than time, so it was not thought odd to represent Christ's birth, his first bath given by two midwives, and the annunciation to the shepherds all in the same picture, as in Duccio's *The Nativity* (c. 1310).

Prominent wave-like or continuum aspects appear in the products of the "Op Art" school (for example, the works of Frank Stella in the 1960's). These works exhibit patterns having a machine-like quality which can be imagined to extend to infinity, dazzling the eye. The "New York" school of artists in the 1950's—Ad Reinhardt, Barnett Newman, Mark Rothko—also produced paintings with continuum properties. Their canvases sport a few different colors arranged in large continuous areas.

The peculiar melting watches in Dali's painting *The Disintegration of the Persistence of Memory* (1952–1954) convey the intellectual message that genetic transmission of characteristics (as symbolized by a fish) is a more permanent form of memory than man-made objects, such as watches. He depicts the surface of the water as a sheet which may be lifted by its corner. The painting has echoes of Einstein's Theory of Special Relativity in its treatment of space and time, but these are not the qualities for which one usually buys a painting. It will ultimately be judged not on its programmatic content but on its artistic merits.

Eadweard Muybridge and High-Speed Photography

Is photography an art? This question was answered negatively in a petition to the French Imperial Court in 1862 as follows:

> Whereas photography consists of a series of completely manual operations which no doubt require some skill in the manipulations involved, but never resulting in works which could *in any circumstance* ever be compared with those works which are the fruits of intelligence and the study of art—on these grounds, the undersigned artists protest against any comparison which might be made between photography and art.

Among the signers were Ingres and Philippe Rousseau. Ironically, many of Ingres' portraits contain clues that he might have clandestinely used daguerreotypes. He posed several of his subjects with hand to head, typical of photographic subjects at the time, who needed to steady their heads during the long exposures of several minutes then required. Also some peculiar left-right reversals occurred between studies for several paintings and the final versions. Lateral transposition is characteristic of this kind of photographic technique, but not of portrait painting unless it is a self-portrait, commonly painted from a mirror image.

The occasion for the rebuke to artistic aspirations on the part of photographers was the famous Mayer and Pierson case (1861–1862) as related by Aaron Scharf in his book, *Art and Photography*. Mayer and Pierson were photographers who claimed that their photographs of Lord Palmerston (the Prime Minister of England at the time) and Count Cavour (an Italian statesman) had been stolen by another photographic team, and they invoked the protection of the French copyright laws. Unfortunately for them, these laws applied only to the arts, so it seemed necessary to request a court to declare photography an art. The decision of the court went against Mayer and Pierson. Their attorney, M. Marie, prepared an appeal suggesting to the court a definition of art that photography conveniently fulfilled and giving the following glimpse of nineteenth-century aesthetics:

> What then is art? Who will define it? Who will say where it begins and where it ends? Who will say: you may go just so far and no father? I put these questions to philosophers who have dealt with them, and we can read with interest what they have written about art in its different forms.
>
> Art, they say, is beauty, and beauty is truth in its material reality.

If we see truth in photography and if truth in its outward form charms the eye, how then can it fail to be beauty! And if all the characteristics of art are found there, how can it fail to be art! Well! I protest in the name of philosophy.

On the other hand a painter often simply copied nature. M. Marie continued:

He couldn't be happier than if he were able to imitate exactly, and he would be quite convinced that he had produced a work of art by coming as close as possible to this nature which charms him and which he admires.

Is the painter any less of a painter when he reproduces exactly?

Arguing in the same vein, M. Marie said that an artist's eye is like a camera, while his hand records nature as a chemical fixes film; likewise a photographer must first conceive of a picture in his mind and then photographically record it. These arguments carried the day, and the court reversed itself in favor of the plaintiffs.

Of the myriad ways in which the new art affected painting, one of the most fascinating was a new way of perceiving motion through instantaneous photography. Traditionally, artists have used three ways to depict motion. The first way is *potential.* A sphere resting on a plane has no potential for motion, but a sphere on a hill gives the impression that movement is about to take place.

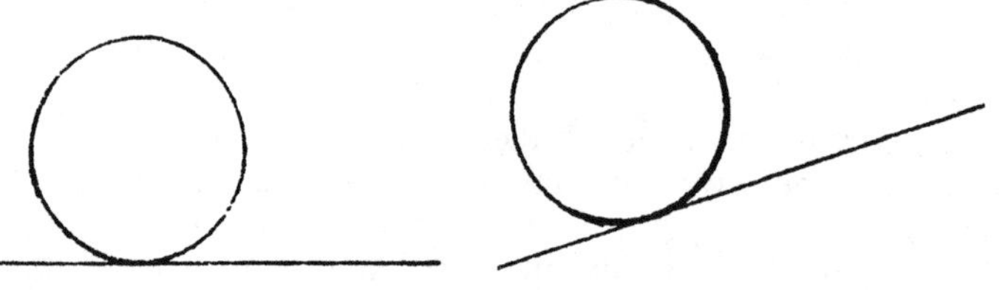

Movement has been arrested as in a "flash" photograph. Experiments in the 1850's had shown that the light from an electric spark stopped the motion of a revolving disc. This demonstration was limited to the laboratory, and a practical method remained to be developed. The second way of depicting motion is *distortion.* In this mode motion is smeared out as in time-exposure photography. Successive stages in the motion blend into a continuous stream. If the artist were showing windshield wipers on an older car, they might be blurred out as shown below.

The third way is *successive stages in frozen motion*, corresponding in photographic terms to a stroboscopic picture or multiple flashes. If we could obtain the sucessive stages in the motion of the windshield wipers, they would look something like the following:

Artists seem to have utilized the latter mode in their earliest attempts to represent motion, for example the multiple horse limbs in some Egyptian temples and tombs commissioned by Rameses II (1250 B.C.), most notably at the Temple of Luxor and at Abu Simbel. The drawings of Leonardo da Vinci record successive stages in motion about as accurately as the unaided human eye can perceive, but technology surpassed this, giving us faster glimpses of motion. Eadweard Muybridge devised the first high-speed photographic technique utilizing wet colloidon plates and shutter speeds of about 1/500th of a second.

Muybridge seems to have always had a romantic streak. Changing his name from Edward Muggeridge to a more archaic form, he set out from England in 1852 to seek his fortune in the California Gold-Rush Country. He learned photography from the San Franciscan, Carleton E. Watson, who specialized in pictures of Yosemite Valley. If one makes allowance for the very high contrast and limited red sensitivity of the wet-plate process, Muybridge's landscapes compare favorably with the present-day work of Ansel Adams. In 1868, the U.S. Government took notice of his work and made him Director of Photographic Surveys. In this capacity he took pictures of the newly-purchased territories of Alaska, Wyoming, Montana, and the West Coast in general.

Hearing of Muybridge's reputation as a photographer, Leland Stanford, President of both the Central Pacific Railroad and the Pacific Mail Steamship Company and ex-Governor of California, telegraphed a request that Muybridge settle an argument with incontrovertible photographic evidence. The question was whether or not during a horse's trot all four legs were off the ground at the same time. The popular opinion was that a horse could not possibly be airborne at any time during a trot. Using a wooden shutter with an eighth-inch slit in it, Muybridge achieved a shutter speed of 1/500th of a second when the shutter was driven by a spring. These first photographs of 1872 have unfortunately been lost, but they were evidently clear enough to

Fig. 55. Consecutive series photographs by Eadweard Muybridge showing phases of a horse's gallop. Stanford U. Library.

satisfy Stanford and were probably the basis for the Currier and Ives 1873 lithograph of Stanford's horse, Occident. Later, in a series of ingenious photographs taken in 1877 and 1878 Muybridge succeeded in capturing successive stages not only of a horse's trot, but also of the gallop and several other gaits. He had devised an improved shutter consisting of double blades which moved in a scissors action; the shutter was tripped electrically from wires on the track. In the case of a trotting horse the wheels of a sulky made electrical contact with the wires, while threads were stretched across the track which a running horse would break as it went by. The threads led to twelve identical cameras whose shutter mechanisms were tripped successively, taking twelve pictures at right angles to the horse's motion. Muybridge calculated his shutter speed at 1/2000th of a second, fast enough to obtain sharp pictures clearly indicating that all four of a horse's legs were off the ground not only during the trot, but also during the canter and gallop.

To suppose that the "photofinish" of a modern horse race is a Muybridge-type photograph is a common error; actually, it corresponds to the *second* way of depicting motion. Muybridge's photographs represent a single instant of time (or nearly so) at several different places within the field of view of the camera lens. In contrast the photofinish camera records several instants of time at a single place—a very odd photograph indeed. A photofinish camera "sees" only a narrow slit. The film is set in motion when the horses are about to cross the finish line, and the image of the slit is recorded on the moving film. Later an artist draws in the finish line. A Muybridge-type photo would tell us which horse won the race, but nothing about place and show, because horses number two and three might have changed places in the race after number one crossed the finish line.

Imagine the sensation created by the publication of the Muybridge pictures. Many intermediate positions of the horse's legs appeared awkward, not at all like the conventional representations of the flying gallop picturing the animal with limbs stretched out fore and aft as in Gericault's *Course de Chevaux à Epsom, le Derby en 1821.* The scientific pictures showed that when the horse was off the ground the legs were jack-knifed underneath the animal. Among many artists who applied the new knowledge in their work were Thomas Eakins in his painting *The Fairman Rogers Four-in-Hand* (1879) and J.-L.-E. Meissonier's revision of his painting *Friedland* when he copied it in water color in 1888.

But not all artists were convinced, a notable exception being Rodin. His sculpture *St. John the Baptist* (1878) depicts a man with both feet planted firmly on the ground, yet it conveys a strong sense that the man is striding forward. This contradicts the high-speed photographs which show a man poised on one foot during a walk. When Rodin was asked how he reconciled the truth of Muybridge's photographs with the artist's claim to copy nature sincerely, he replied that art is more truthful than photography, because the artist produces the impression of a movement which takes several moments for accomplishment, while in photography the abrupt suspension of time is contrary to reality.

Another effective criticism based on the phenomenon of persistence of vision came from the American scientist Ogden Rood.

> [This explains] the transparency of rapidly moving wheels. Owing to the same cause, the limbs of animals in swift motion are only visible in a periodic way, or at those moments when their motion is being reversed; during the rest of the time they are practically invisible. These moments of comparative rest are seized by artists for delineation, while the less discriminating photograph is apt to reproduce intermediate positions, and thus produce an effect which, even if quite faithful, still appears absurd.

Muybridge responded by saying that an arbitrary symbol may become associated with an existing fact, and that this association may be reinforced at various stages in life until it is extremely difficult to dissociate them, even when reason convinces us that they have no true relationship.

> So it is with the conventional galloping horse; we have become so accustomed to see it in art that it has imperceptibly dominated our understanding, and we think the representation to be unimpeachable, until we throw all our preconceived impressions on one side, and seek the truth by independent observation from Nature herself. During the past few years the artist has become convinced that this definition of the horse's gallop does not harmonize with his own unbiased impression, and he is making rapid progress in his efforts to sweep away prejudice, and effect the complete reform that is gradually but surely coming.

This has also been the cry of scientists with revolutionary ideas from Galileo to Schrödinger and Heisenberg, the co-discoverers of quantum mechanics. To think that the sun might have spots on its supposedly perfect surface violated the common sense of Galileo's contemporaries just as it astounded the public in the 1920's to contemplate matter waves. Since the human mind is remarkably adaptable, by now our common sense has assimilated these wonders and is ready to react to any new theory which might present itself.

Does the Muybridge high-speed photograph represent "the truth," that is "what we would have seen if we had been there?" It would be if we observed at one instant of time (which we cannot do) from one particular vantage point (that of the camera lens) with our head immobile, if we closed one eye, if we saw with the equivalent of a 150 mm lens, if we saw with the equivalent of Tri-X film developed in D-76 and printed on Kodabromide paper and so on. So the Muybridge way of making photographs is just as artificial as the photofinish camera technique. Neither pictures reality as we humans perceive it.

Ironically, the Muybridge instantaneous photographs in all their definition and clarity prepared the way for the beginning of Impressionism in art. How did that come about? The blurred image as a natural, human response gained acceptance because of the study of persistence of vision phenomena by physiologists such as Rood. So when Impressionism began at the end of the nineteenth century, the public was in a more receptive mood.

Leonardo da Vinci

The variety of different fields in which Leonardo made significant advances is mind-boggling: engineering, painting, geology, paleontology, drawing, botany, zoology, physics, anatomy, philosophy and writing. Obviously anything less than a book-length study can only be a superficial appraisal of this fifteenth-century genius. To recount the instances in which his career impinges topics previously discussed in this book virtually constitutes a review.

Leonardo's Creative Process. His basic approach to any problem seems to have been primarily visual, either making a drawing or model. He said that the eye was the window of the soul and the chief means whereby the understanding can appreciate the infinite works of nature. Compare this mode of understanding with Albert Einstein's, which he discussed in a letter to Jaques Hadamard as follows:

> The words or language, as they are written or spoken, do not seem to play any role in my mechanism of thought. The psychical entities which seem to serve as elements in thought are certain signs and more or less clear *images* which can be "voluntarily" reproduced and combined.
>
> The above mentioned elements are, in my case, of *visual* and some of muscular type. Conventional words or other signs have to be sought for laboriously only in a secondary stage, when the mentioned associative play is sufficiently established and can be reproduced at will. [emphasis added]

In the terminology of Chapter 1, both Leonardo and Einstein took advantage of the special qualities of the brain's *right* hemisphere as they grappled with challenging problems.

A single page of Leonardo's notebooks often reveals hidden consistency beneath an apparently disorderly assortment of sketches, mirror writing, and mathematical fragments. Kenneth Keele cites an example of a page (Windsor Collection, #12283) which contains drawings of grasses curling around an arum lily, cumulus clouds, rippling waves, trees, a screw press, segments of circles, an old man with curly hair, and a note about the preparation of curly hair—a holistic approach to the general concept of curvature. Such jottings are difficult to study because Leonardo made no effort to organize his thought and often returned to the same subject years later in another notebook.

Phrases recalling Poincaré's unexpected illuminations crop up in his writings on mathematics. Leonardo mentions, for example, a mathematical invention which came to him "as a gift on Christmas morning 1504;" another discovery came to him on the night of November 30, 1504 as the hour, light, and page were all coming to a close. Such instances give no more than a glimpse of his thought processes.

Leonardo and Newton. In several respects the researches of Leonardo anticipate Newton. Leonardo observed and carefully described the colors in the rainbow, but he wondered whether the colors were subjective ones generated in the eye of the beholder. To test this he viewed the rainbow through a water-filled glass and found the colors of the rainbow in each bubble of coarse glass which acted as a spherical lens. In addition he placed the glass on a window sill, allowed the sun's rays to strike it from behind, and observed colored patterns on the pavement at his feet. Newton went further than this, experimenting with a shaped beam of light from a slit and a glass prism which separated the light into a broad spectrum. Not stopping there, he tested each individual colored ray to see if it could be further refracted and also successfully recombined the colored rays into white light again.

In his writings Leonardo touched upon two of Newton's laws of motion, the first and the third. He believed, as did Aristotle, that a body at rest persists in that state indefinitely until disturbed. But he did not believe in a vacuum, and therefore could not conceive of motion that would persist forever unimpeded by friction. He assumed, as did his contemporaries, that motion would cease when the "impression of the motive force" in a body was exhausted. He came nearer to the Newtonian conception of the law of action and reaction with the following statement: "An object offers as much resistance to the air as the air does to the object." (Codex Atlanticus, fol. 381v-a). In another place he truly rose to Newtonian generality. "Every body goes in a direction opposite to the place from which it is driven by the object that strikes it . . .The body strikes the object in the same measure as the object strikes the said body." (Codex Arundel, 263, fol. 135r).

Architecture and City-Planning. Leonardo's concept of a two-level city seems far-sighted enough for the needs of the nineteenth or twentieth century. Heavy traffic consisting of carts, wagons, and beasts of burden was restricted to the lower level, while the upper level was reserved for pedestrians and light traffic, and included an arcade where people might walk sheltered from the elements. The streets were to be sloped from the center to the sides to provide drainage. Openings in the upper level provided light to the lower level, and the two levels were connected by circular stairs. Access to the upper level was also provided by ramps outside the city walls. Included in the plan were limits for heights of buildings, an adequate water supply, and ample provision for a sewage system with main trunk, laterals, and individual house connections.

Leonardo and Huyghens. Huyghens' sophistication in designing escapements and pendulum suspensions was certainly far beyond Leonardo's abilities. The means of measuring time available to Leonardo were no different from those available to Galileo: the heart rate, musical tempi, sandglasses, and water clocks. But in certain other respects Leonardo seems rather close to the discoveries of Huyghens. Leonardo's diagram of the intersection of waves produced by two stones dropped into a pond (Ms. A, fol. 61r) is strongly reminiscent of the fourth figure in Huyghens' *Traité de la Lumiere* in which expanding circles of light originate from different parts of a candle flame, the basis for

Huyghens' construction in optics. To discover why the expanding wave circles could pass through rather than dash against each other, Leonardo placed bits of straw on the surface of the water. The straws merely bobbed up and down as the wave passed under them, a behavior characteristic of transverse waves, not longitudinal. By analogy he extended the concept to light. "Just as a stone flung into water becomes the center and cause of many circles . . . so any object placed in the luminous atmosphere diffuses itself in circles and fills the surrounding air with images of itself." (Ms. A, fol. 9r). Huyghens did not have any notion of the transverse nature of the light waves; the concept of the transverse and vibratory nature of light was introduced later by Thomas Young. Leonardo's work paralleled Huyghens in another area. Leonardo designed several machines for grinding concave metal mirrors; Huyghens designed similar machines for making glass lenses.

The Observant Eye. Before the development of high-speed photography discussed earlier in this chapter, the motion of living creatures could only be recorded if the artist had an acute perception of intermediate stages of motion, a good visual memory, and an exceptionally fast hand. Leonardo's drawings of animals and men in motion probably represent the limit of human perception in this regard. Consider, for example, his sketches of the flight of birds (Ms. E), the positions of cats (Windsor Col. 12363), a struggle between a horse and a dragon (Windsor Col. 12331), a man with a sledge hammer (Windsor Col. 12641 *verso*), and men working on various tasks (Windsor Col. 1644-46). That either animals or men could be posed in such extreme positions seems unlikely.

If Leonardo undertook most of the above studies and his work in anatomy to improve his painting, as is commonly supposed, then he must have become fascinated with the studies as ends in themselves, because the number of his surviving paintings is so small. Only about a dozen authenticated Leonardo paintings exist, but for most people these are the supreme examples of his art.

The Engines of War. Until the latter part of the nineteenth century, when his notebooks began to be widely published, Leonardo was primarily revered for his painting, but he puts his ability as painter *last* in his letter (1481) of application to Lodovico Sforza, then Regent and later Duke of Milan. He begins with a long list of military engineering talents: the ability to make bridges, trellises, scaling ladders, tunnels, cannon, mortars, field pieces, dart throwers, and fire throwers. He concludes with some of his civil accomplishments: architecture for public edifices or private dwellings, water distribution systems, sculpture in marble, bronze or terra cotta, and lastly painting. After the French captured the Duke, Leonardo was employed as chief inspector of military fortifications and military engineer for Cesare Borgia in th Romagna. He returned to Florence in 1506 and began his portrait *Mona Lisa*.

Viewing Leonardo's military drawings is like a trip through the Wax Museum: one sees all the machines designed for killing and maiming human beings. He sketches a variety of cleverly designed weapons: a rapid-fire catapult, a giant crossbow stretched by means of a screw mechanism, a breech-loading siege gun, machine guns, a field piece with three breech-loading barrels, mortars discharging exploding shells and shrapnel, mortars firing barrages, and horizontal wheels with rotating clubs or scythes. No record exists of his ever regretting having undertaken this military work, no soul-searching in the manner of Oppenheimer and Einstein. If he apparently did not examine his conscience on these matters, neither did he devise means whereby the entire human race could be brought to the verge of extinction.

Bibliography

Anonymous, *Leonardo da Vinci* (Reynal, New York, 1956) from the great Leonardo exposition in Milan 1938.

Clark, Kenneth, *The Drawings of Leonardo da Vinci in the Collection of Her Majesty the Queen at Windsor Castle* (Phaidon, London, 2nd edition 1969) Vols. 1-3.

Dali, Salvadore, *Dali on Modern Art,* translated by Haakon M. Chevalier (The Dial Press, New York, 1957).

Escher, Maurits C., *The Graphic Work of M.C. Escher,* translated by Brigham, J.E. (Hawthorne, New York, 1960).

Gardner, Martin, "The Eerie Mathematical Art of Maurits C. Escher," Scientific American **215**, 110, April 1966.

Haas, Robert B., *Muybridge, Man in Motion* (Univ. of Calif. Press, Berkeley, 1976).

Hammacher, A.M., *Renê Magritte,* translated by Brockway, J. (H.N. Abrams, New York, 1974).

Hartlaub, G.F., *Zauber des Spiegels* (R. Piper and Co., Munich, 1951).

Hendricks, Gordon, *Eadweard Muybridge* (Grossman, New York, 1975).

MacDonnell, Kevin, *Eadweard Muybridge: the Man who Invented the Moving Picture* (Little, Brown and Co., Boston, 1972).

Nicolson, Benedict, *Joseph Wright of Derby,* Vols. 1 and 2, (Routledge and Kegan Paul, London, 1968).

Parsons, William B, *Engineers and Engineering in the Renaissance* (M.I.T. Press, Cambridge, 1968) Reprinted from 1939 edition.

Pennsylvania University, *Animal Locomotion: The Muybridge Work at the University of Pennsylvania,*(Arno Press, New York, Reprint Edition 1973).

Scharf, Aaron, *Art and Photography* (Penguin Books, Harmondsworth, England, 1968).
Scharwz, Heinrich, "The Mirror of the Artist and the Mirror of the Devout," in *Studies in the History of Art Dedicated to William E. Suida on his Eightieth Birthday* (Phaidon Press, London, 1959).
____, "The Mirror in Art," The Art Quarterly **15**, 97, Summer 1952.
Shurlock, F.W.,"The Scientific Pictures of Joseph Wright," Science Progress **17**, 432, 1923.
Stanford University Department of Art, *Eadweard Muybridge: The Stanford Years, 1872-1882* (Stanford University, 1972).
Teuber, Marrianne L.,, "Sources of Ambiguity in the Prints of Maurits C. Escher," Sci. Am. **230**, 90, July, 1974.
Trevor-Roper, Patrick, *The World Through Blunted Sight* (Bobbs-Merrill Co., Indianapolis, 1970).

Index